Fundamental Mathematics 2:

Tests and Further Exercises

G.R. Gilbart-Smith,
R.Q. Edmonds and
D.J. Hillard

Hodder & Stoughton
LONDON SYDNEY AUCKLAND TORONTO

First published 1986
Fifth impression 1991

British Library Cataloguing in Publication Data
Gilbart-Smith, G. R.
Fundamental mathematics
2 : Tests and further exercises
1. Mathematics—Examinations, questions, etc.
I. Title II. Edmonds, R. Q. III. Hillard, D. J.
510′.76 QA43
ISBN 0 7131 8413 2

Printed and bound in Great Britain for the educational publishing division of Hodder & Stoughton Limited, Mill Road, Dunton Green, Sevenoaks, Kent by Thomson Litho Ltd, East Kilbride, Scotland.

Contents

1 Common Entrance Paper I: 30-minute tests 1

2 Common Entrance Paper II: revision topics
2.1 Sets and Venn diagrams 13
2.2 Perimeter, area and volume 17
2.3 Averages 21
2.4 Speed 26
2.5 Algebra problems 30
2.6 Angle calculations 34
2.7 Transformations 38
2.8 Ratio and percentage 46

3 Common Entrance Paper II: 30-minute tests 49

Preface

Our first volume, *Fundamental Mathematics: Problems and Exercises*, aims to lay a strong foundation on which to build in the future. Though its content is based on the Common Entrance syllabus, it is not exclusively intended for use in Independent Schools. Since its publication, many users have indicated the need for a book which assists Common Entrance candidates with what can be a daunting run-up to their examination. In particular, many pupils who are quite confident on short questions find themselves confused by the longer type of question they encounter in Paper II. This book is finely tuned to the needs of such candidates.

The first chapter consists of twelve tests in the style of Paper I. Each test is designed to last 30 minutes. Each pair of tests (1 and 2, 3 and 4, etc.) may be combined into a full trial paper, if desired. Whether used in this way or not, their broad coverage of the syllabus will give valuable material for revision. The second chapter concentrates on parts of the syllabus which most need extra practice for Paper II. Without going beyond the scope of the *material* introduced in the first volume, it aims to build up confidence for Paper II by following the *style* of its questions. Each exercise is carefully graded to introduce the more diffident pupils to the style, but the more able candidates will find plenty to keep them interested and occupied towards the end of the exercise. The final chapter of the book comprises twelve tests in the style of Paper II, designed in pairs like those in chapter 1.

It is hoped that pupils who are only taking the compulsory papers (and their teachers!) will find that this volume takes something of the strain off the final period of preparation for the examination, and enables the pupils to sit the exam in a relaxed frame of mind while showing their potential to full advantage. Teachers may like to use this book in parallel with the first volume during the final year before Common Entrance or may prefer to time their presentation of the first volume to finish with a few months to spare, in order to use this book alone thereafter. Used in either way it will prove an indispensable companion to the first volume.

More able pupils should finish the first volume with a year or more to spare before taking Common Entrance (including Mathematics Paper III) or Scholarship examinations. This book will prove a useful bridge between Books 1 and 3 for such pupils, introducing them to problems which involve several stages of thought for their solution. Teachers may, however, prefer to use this book in conjunction with Book 3 to provide revision material on the 'basics' while pupils study the more advanced topics.

For the inspiration to produce this volume and the accompanying Book 3, we are greatly indebted to our many colleagues throughout the country who have encouraged us. Our special thanks are due to the late Mr. John Bickmore of Yardley Court School for the assistance he gave the project at an early stage. Having goaded us on to get our first volume to press, we hope he would have been pleased to see it develop into a series.

G.R. Gilbart-Smith
R.Q. Edmonds
D.J. Hillard
January, 1986

1

Common Entrance Paper I: 30-minute tests

Within each exercise, a possible mark scheme is indicated in parentheses.

Exercise 1.1

1. Calculate **(a)** 3.4 − 1.86 *(2)* **(b)** 3.567 ÷ 4.1 *(5)*
2. Using index notation, write 360 as the product of prime factors. *(3)*
3. Robert and Sarah share a box of toffees in the ratio 3 : 4. If Robert's share is 18 toffees, how many are there in the box altogether? *(3)*
4. Add together 3.4 m + 370 cm + 450 mm.
   Give your answer in **(a)** cm *(4)* **(b)** m *(1)*
5. Oranges cost 18 p each. Mr Haydon has £5 to spend. How many oranges can he buy and how much change will he receive? *(4)*
6. Simplify **(a)** $6a^2 - 3a + a^2$ *(2)* **(b)** $3a^2b \times 2a^3b^2$ *(2)*
   (c) $3a^2b^3 \div 6a^3b$ *(3)*
7. If $a = 3$, $b = -2$ and $c = 1$, find the value of
   (a) $3a^2b - 2bc$ *(4)*
   (b) $(2ab)^2$ *(2)*
8. A rectangle is L cm long and W cm wide. To find the perimeter, P cm, the sum of the length and the width is doubled.
   (a) Write down a formula for P in terms of L and W. *(3)*
   (b) Make L the subject of the formula. *(2)*
9. The graph (Figure 1) shows the number of cornets, wafers, and choc-ices sold at Benny's ice-cream van on a day last summer. A cornet costs 20 p, a wafer 15 p and a choc-ice 25 p. What were the total takings for the day? *(7)*

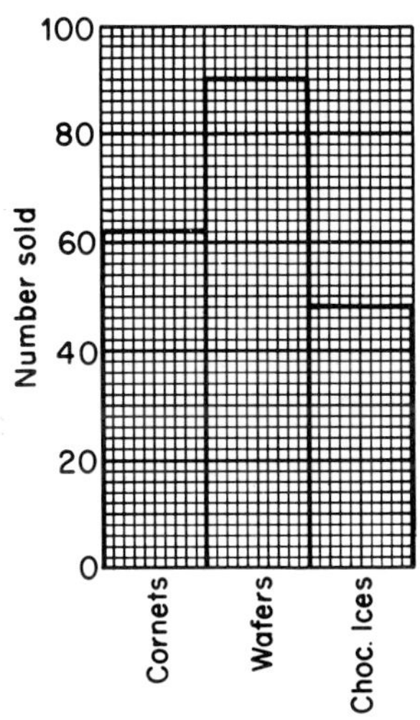

Figure 1

10. Calculate the sum of the interior angles of a pentagon. *(3)*

Exercise 1.2

1. Calculate (**a**) $3\frac{1}{2} + 1\frac{2}{3}$ (*3*) (**b**) $3\frac{1}{2} - 1\frac{2}{3}$ (*3*)
2. (**a**) Estimate the value of $\dfrac{63.6 \times 4.2}{8.9}$, correct to 1 significant figure. (*2*)
   (**b**) Write 0.0043 in standard form. (*2*)
3. (**a**) What is 120 as a percentage of 480? (*2*)
   (**b**) Frances needs 65% to pass her Mathematics exam, which is marked out of 180. What is the least mark Frances must obtain to pass? (*2*)
4. (**a**) What is the width of a rectangle which is 21 cm long and has an area of 168 cm²? (*3*)
   (**b**) What is the area of a square whose sides are 2.5 m long? (*3*)
5. Michael runs 40 km in 2 hours 30 minutes. What is his average speed? (*4*)
6. Factorise $3x^2y - 6xy^4$. (*3*)
7. Solve (**a**) $x + 4 < 2x$ (*2*) (**b**) $\frac{1}{5}y + \frac{1}{2} = 3$ (*4*)
8. (**a**)

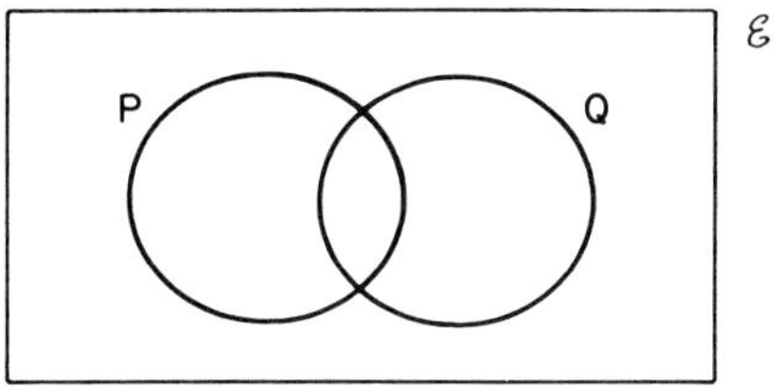

Figure 1

Copy Figure 1 and shade in $P \cap Q'$. (*2*)

(**b**)

Figure 2

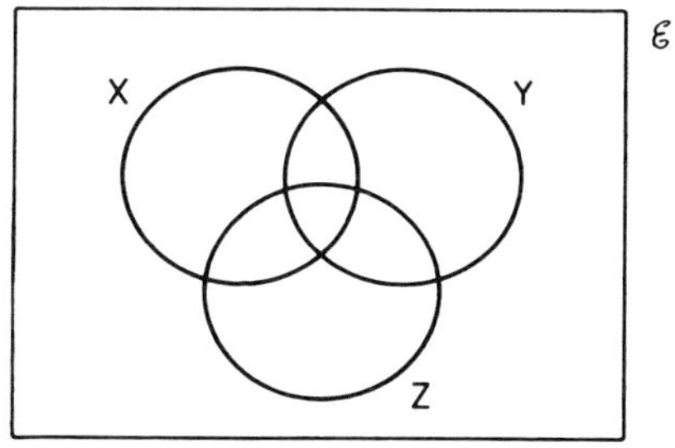

Copy Figure 2 and shade in $(X \cap Y) \cup Z$. (*2*)

9. In Figure 3:
   AC is parallel to EG,
   ED = EF,
   $D\hat{F}G = 114°$.
   Calculate, giving reasons,
   (**a**) $D\hat{F}E$ (*2*)
   (**b**) $F\hat{E}B$ (*2*)
   (**c**) $C\hat{B}E$. (*2*)

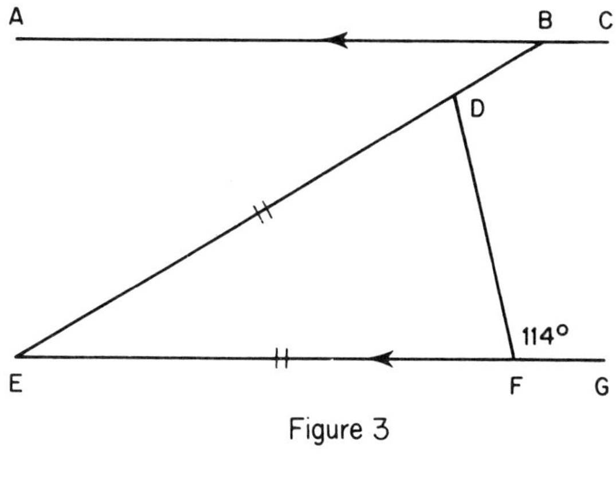

Figure 3

10. (**a**) Mark out x and y axes from 0 to 8. On these axes plot points A(1, 1), B(3, 1) and C(3, 3), and join them to form triangle ABC. (*1*)
    (**b**) Triangle ABC is rotated through 90° clockwise about (4, 3) to give triangle A′B′C′. Plot triangle A′B′C′. (*2*)
    (**c**) Triangle A′B′C′ is reflected in the line $x = 5$ to give triangle PQR. Plot triangle PQR. (*2*)
    (**d**) What is the vector of the translation mapping triangle ABC onto triangle RQP? (*2*)

Exercise 1.3

1. **(a)** Calculate $3.78 + 16 + 1.4$ *(2)*
   (b) **(i)** Calculate 2.48×0.36 *(4)*
   (ii) Write down the answer to 248×3.6 *(1)*
2. **(a)** Express **(i)** 42 **(ii)** 63 as the product of prime factors. *(4)*
   (b) Hence find **(i)** the HCF of 42 and 63 *(2)*
   (ii) the LCM of 42 and 63. *(2)*
3. Four men have mass 75 kg, 83 kg, 87 kg, and 90 kg. What is their mean (average) mass? *(3)*
4. **(a)** Add in base 6:

$$\begin{array}{r} 234 \\ 15 \\ 533 \\ \hline \\ \hline \end{array}$$

(2)

   (b) Subtract in binary:

$$\begin{array}{r} 10110 \\ 1011 \\ \hline \\ \hline \end{array}$$

(3)

5. Find the volume of a cuboid which measures 30 cm by 20 cm by 12.5 cm. *(3)*
6. Remove the brackets and simplify $3(2a + b) - 2(a - b)$. *(4)*
7. If $p = 3$ and $q = -2$, find the value of $(pq)^2 - 2p^2q$. *(4)*
8. A pencil costs x pence and a pen costs $(x + 10)$ pence.
   (a) What is the cost of 3 pencils and 4 pens? *(2)*
   (b) Simplify your answer to part (a). *(1)*
   (c) If the total cost is £1.10, what is the cost of a pencil? *(3)*
9. Each interior angle of a regular polygon is 140°. How many sides does it have? *(3)*
10.

Figure 1

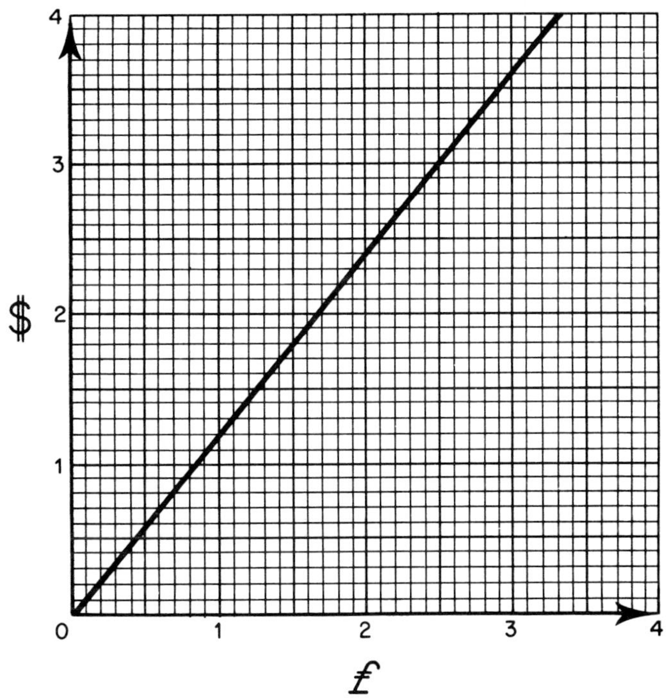

The graph (Figure 1) shows the rate of exchange between the pound (£) and the dollar ($).

(a) What is the equivalent of **(i)** £1.50 in $ *(2)* **(ii)** $2.70 in £? *(2)*
(b) A breakfast costs £2.90 in London and $3.00 in New York. In which city is it cheaper to buy the breakfast and by how much? Give your answer in cents. *(3)*

Exercise 1.4

1. Calculate (**a**) $7\frac{1}{2} \times 1\frac{4}{5}$ *(3)* (**b**) $7\frac{1}{2} \div 1\frac{4}{5}$ *(2)*
2. Write 19.86 (**a**) correct to 2 significant figures *(1)*
   (**b**) correct to 1 decimal place *(1)*
   (**c**) in standard form. *(2)*
3. (**a**) What is $7\frac{1}{2}$% of £1400? *(3)*
   (**b**) David gained 27 out of 60 in a Maths test. What percentage is this? *(3)*
4. What is the area of the room shown in Figure 1,
   (**a**) in m^2 *(4)*
   (**b**) in cm^2? *(2)*

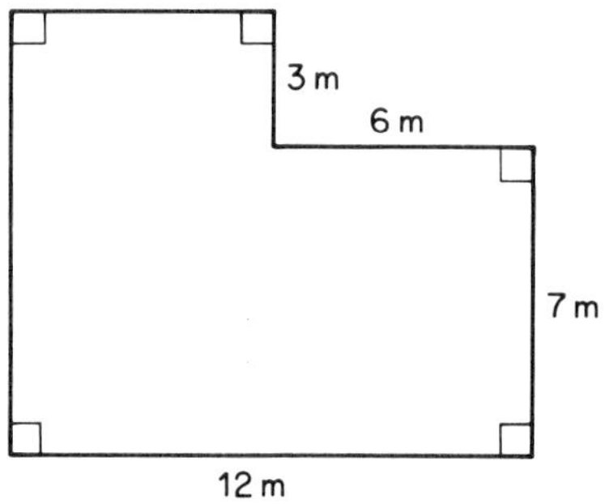

Figure 1

5. A tortoise travels at a speed of 40 m/h. How far does it travel in 2 hours 45 minutes? *(3)*
6. Simplify $\dfrac{6ab^2 - 4a^3b^8}{2ab^2}$. *(3)*
7. Solve (**a**) $2(x + 2) = 3(x - 2)$ *(3)* (**b**) $\dfrac{x+2}{6} = \dfrac{x-1}{4}$. *(3)*
8. $\mathscr{E}$ = {positive whole numbers < 10}
   A = {multiples of 2}
   B = {multiples of 3}
   (**a**) Copy the diagram (Figure 2) and place each member of $\mathscr{E}$ in the correct region. *(3)*
   (**b**) What are the members of $A \cap B'$? *(2)*

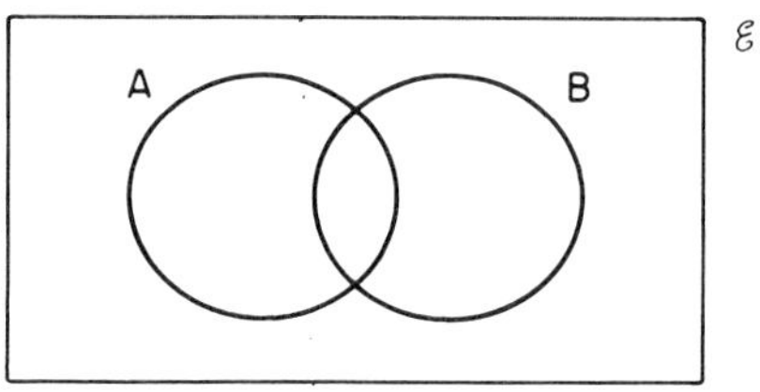

Figure 2

9. In Figure 3:
   O is the centre of the circle,
   OC is parallel to AB,
   $A\hat{O}B = 124°$.
   Calculate, without giving reasons:
   (**a**) $O\hat{A}B$ *(2)*
   (**b**) $B\hat{O}C$ *(2)*

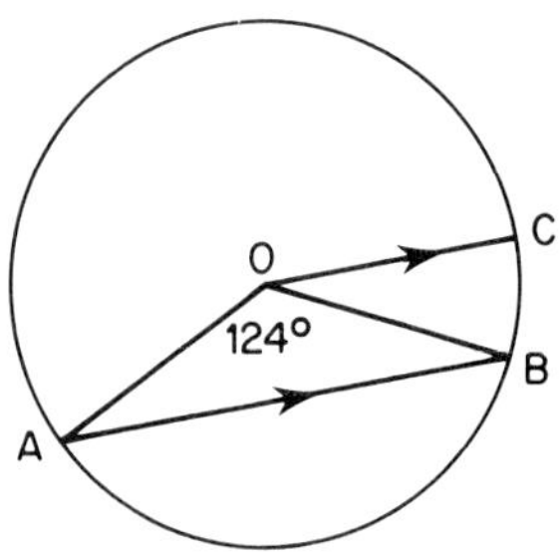

Figure 3

10. (**a**) Mark x and y axes from 0 to 8. On these axes, plot and join P(2, 7), Q(1, 4) and R(3, 4). *(1)*
    (**b**) What is the equation of the axis of symmetry of △PQR? *(2)*
    (**c**) Reflect △PQR in $y = x$. Call the image △P′Q′R′. *(3)*
    (**d**) Rotate △P′Q′R′ through −90° about P′. Call this image △P′Q″R″. *(2)*

Exercise 1.5

1. Calculate (**a**) $6.37 - 1.9 + 2.36$ *(2)* (**b**) $(0.3)^3$ *(3)*
2. By using prime factors, find the square root of 441. *(4)*
3. The perimeter of a triangle is 36 cm. The sides are in the ratio 2 : 3 : 4. What is the length of the longest side? *(3)*
4. A kilogram of strawberries costs £1.54. What is the cost of 2.5 kilograms? *(4)*
5. Calculate (**a**) $234_7 \times 45_7$, working in base 7 *(3)*
   (**b**) $302_6 \div 5$, working in base 6. *(2)*
6. Simplify (**a**) $7ax - 4ay - 5ay - 3ax$ *(2)* (**b**) $\dfrac{(2a^3b)^2}{4a^2b}$ *(3)*
7. If $x = \frac{1}{2}$ and $y = -6$, find the value of $8x^2 - 4xy$. *(4)*
8. The total cost (C) of hiring a fishing boat is a payment of £20 plus £5 for each hour of use.
   (**a**) Write down a formula for C after h hours of use. *(3)*
   (**b**) Make h the subject of the formula. *(2)*
9. In Figure 1, O is the centre of the regular polygon ABCDE.
   Calculate, giving reasons:
   (**a**) $A\hat{O}B$ *(3)*
   (**b**) $O\hat{B}A$ *(3)*
   (**c**) $E\hat{D}A$. *(3)*

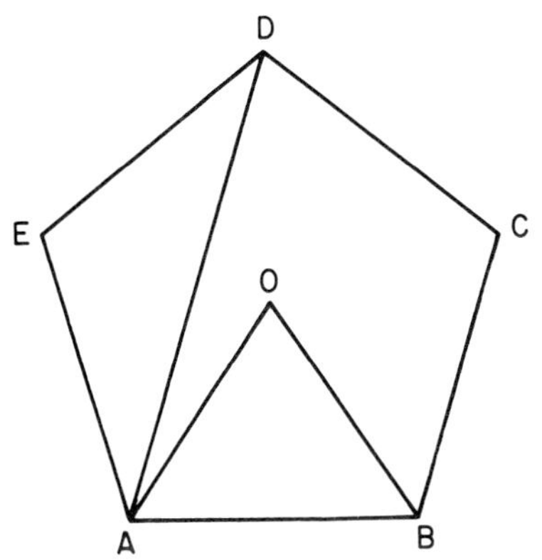

Figure 1

10. Figure 2 shows John's afternoon walk, during which he visits St. David's Church before returning home.
    (**a**) How far is the church from John's home? *(1)*
    (**b**) How long did John spend visiting the church? *(2)*
    (**c**) At what speed did John travel home? *(3)*

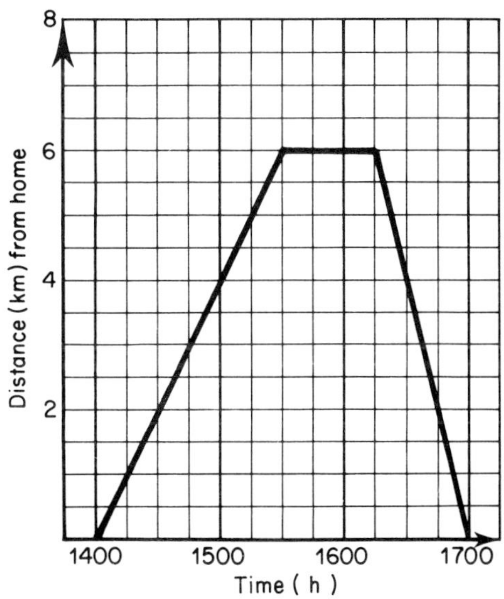

Figure 2

Exercise 1.6

1. Calculate (**a**) $\frac{2}{3}$ of $1\frac{7}{8}$ *(2)* (**b**) $5\frac{1}{3} \div (3 - 1\frac{6}{7})$. *(4)*
2. (**a**) Write $\frac{13}{20}$ as a decimal. *(2)*
   (**b**) Write 0.84 as a fraction in lowest terms. *(2)*
3. A cheese has mass 1 kg. John cuts off a slice of mass 80 grams. What percentage of the cheese is left? *(4)*
4. Calculate the area of the trapezium (Figure 1). *(4)*

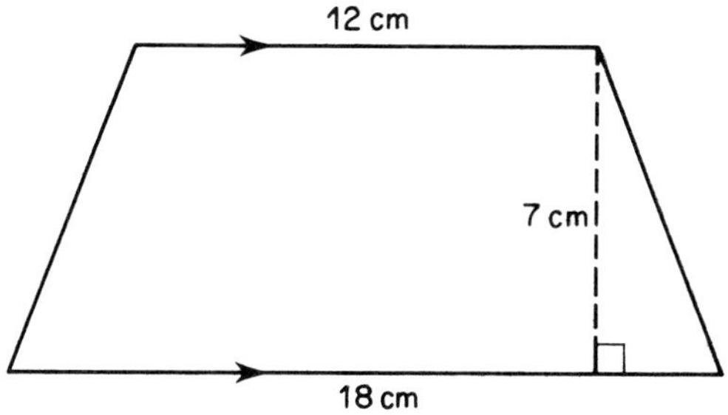

Figure 1

5. A train leaves London at 2245 and arrives in Aberdeen at 0825 the next day. How long does the journey take? *(2)*
6. Solve (**a**) $4 + 2(x - 1) = 5$ *(3)* (**b**) $\frac{x}{3} - 1 < 6$ *(2)*
7. Factorise $8p^3q + 4pq^2$. *(3)*
8.
   (**a**) What is the name given to the polygon ABCDE (Figure 2)? *(1)*
   (**b**) Measure ED. *(1)*
   (**c**) Measure the only acute angle. *(2)*
   (**d**) What kind of angle is $B\hat{C}D$? *(1)*
   (**e**) Measure the size of reflex angle BAE. *(2)*

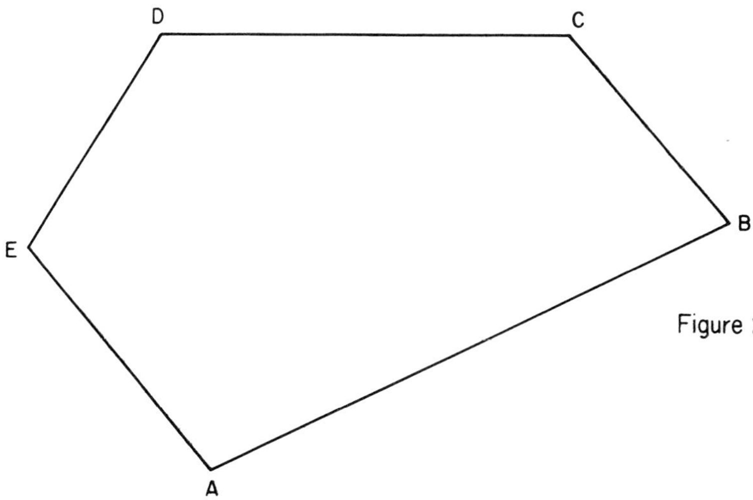

Figure 2

9. Place one of the symbols $\in$, $\cup$, $\cap$, $\subset$ in each of the statements to make it true:
   (a) {2, 4} .... {even numbers} (*1*)
   (b) 7 .... {prime numbers} (*2*)
   (c) {multiples of 4} .... {multiples of 3} = {multiples of 12} (*2*)
10. (a) Mark x and y axes from -3 to 3. On these axes, plot the rectangle which has vertices $(-1, 1)$, $(1, 1)$, $(1, 2)$ and $(-1, 2)$. Label it A. (*2*)
   (b) Translate A by $\begin{pmatrix} 2 \\ -3 \end{pmatrix}$. Label the image B. (*3*)
   (c) Reflect B in $x = 0$. Label the image C. (*3*)
   (d) What are the coordinates of the point about which B can be rotated through 180° to give C? (*2*)

Exercise 1.7

1. Calculate (a) $4.8 - 5.6 + 3.5$ (*3*) (b) $27.8 \div 0.5$ (*4*)
2. (a) Find the LCM of 15 and 18. (*3*)
   (b) Hence calculate $\frac{8}{15} + \frac{5}{18}$. (*2*)
3. The school boiler is working for 15 hours a day. To save fuel, these hours are cut in the ratio of 4 : 5. For how many hours does the boiler now work? (*3*)
4. A train is meant to arrive at 1010. It actually arrives at the following times:

Monday	1012
Tuesday	1007
Wednesday	1015
Thursday	1020
Friday	1005
Saturday	0959
Sunday	1005

   On average, does the train arrive early or late? By how many minutes? (*6*)
5. Write (a) 134_8 as a number in base 10 (denary) (*3*)
   (b) 18_{10} as a number in base 2 (binary). (*2*)
6. Simplify (a) $5 - 3(a - 2)$ (*3*) (b) $\frac{x}{2} - \frac{x}{3}$. (*3*)
7. If $p = 3$ and $q = \frac{1}{2}$, find the value of (a) $4pq^2$ (*3*) (b) $\frac{p}{q}$. (*2*)
8. Make p the subject of the formula $q = 3p - 8$. (*4*)
9. Draw AB = 6 cm. Using ruler and compasses only, construct the mediator of AB.
   The mediator cuts AB at C. Mark C on the figure and construct the locus of points 2 cm from C. (*5*)
10. How many sides does a polygon have, if the sum of the interior angles is 900°? (*4*)

Exercise 1.8

1. Calculate (**a**) $\frac{3}{4}$ of £7.80 (*3*) (**b**) $3\frac{1}{2} - 1\frac{7}{8} + 2\frac{3}{4}$. (*4*)
2. Write 42.36 (**a**) correct to 2 significant figures (*1*) (**b**) correct to 1 decimal place (*1*) (**c**) in standard form. (*2*)
3. Each year, David pays 5% of his salary into a savings account. If he pays in £600 one year, what is his salary? (*4*)
4. What is the perimeter of a square whose area is 144 m^2? (*4*)
5. After a voyage which lasted 6 hours 40 minutes, a ferry docks at 0215. At what time did the ferry start the journey? (*2*)
6. Simplify $\dfrac{4p^2q - 2pq^3}{2pq}$. (*4*)
7. Solve (**a**) $2x - 2 < 7$ (*2*) (**b**) $\frac{1}{2}(x - 3) - \frac{1}{5}(x + 2) = 2$. (*5*)
8. $\mathscr{E}$ = {whole numbers 1–10 inclusive}
   B = {multiples of 2}
   T = {multiples of 3}
   P = {prime numbers} (1 is *not* prime)
   (**a**) Copy the diagram (Figure 1) and place each member of $\mathscr{E}$ in the correct region. (*3*)
   (**b**) What are the members of $B \cap T' \cap P'$? (*1*)
   (**c**) How many members has $(B \cup T)$? (*1*)

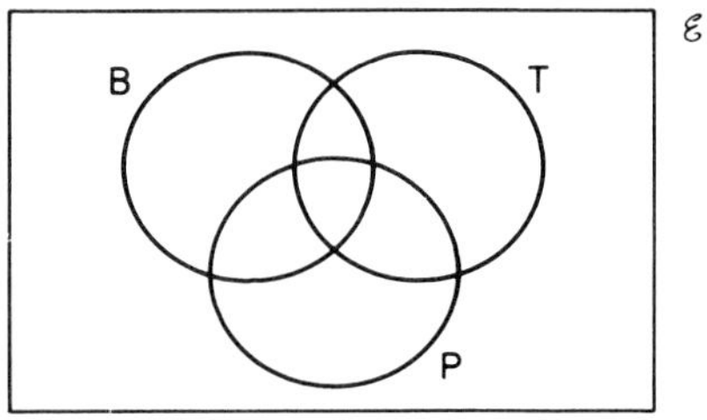

Figure 1

9. In Figure 2:
   ABD is a straight line,
   CA = CB,
   $\hat{ACB} = 40°$.
   (**a**) Which angles are equal? (*1*)
   (**b**) What is the size of $\hat{ABC}$, in terms of x? (*2*)
   (**c**) Form an equation and solve it to find the value of x. (*3*)

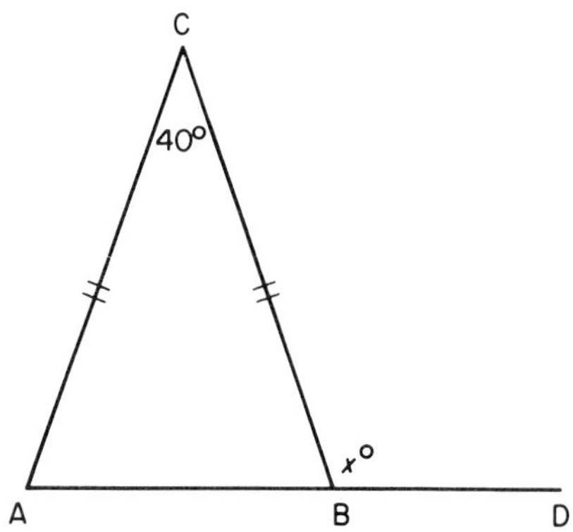

Figure 2

10. (**a**) Mark x and y axes from 0 to 8. On these axes, plot the triangle with vertices (2, 4), (4, 4), (2, 7). Label it A. (*1*)
    (**b**) Rotate △ A through a half turn about the point (5, 4). Label the image B. (*2*)
    (**c**) Reflect △ B in $x = 5$. Label the image C. (*2*)
    (**d**) Draw in △ D to make the complete figure have line symmetry. (*2*)

Exercise 1.9

1. Calculate 4.23×0.0087. (*5*)
2. By using prime factors, find the HCF of 42 and 70. (*5*)
3. On a farm, the ratio of the number of cows to sheep is 3 : 10. If there are 250 sheep, how many cows are there? (*4*)
4. In this question, 13:2 stands for 13 years 2 months. 5 children are aged 13:2, 12:4, 12:8, 13:1 and 12:6. What is their mean (average) age? (*4*)
5. 768 cm^3 of sand is placed in a tin which has a square base of side 8 cm. How deep is the sand? (*4*)
6. Simplify **(a)** $-7y + 3y - 4y$ (*1*) **(b)** $\frac{2a^3b \times 4ab}{6a^2b^3}$. (*3*)
7. If $a = 3$ and $b = -2$, find the value of **(a)** $a^2 - b^2$ (*3*) **(b)** $(a - b)^2$. (*3*)
8. Make p the subject of the formula $x = 3pq + 4$. (*4*)
9. Draw **(a)** a quadrilateral with only 2 axes of symmetry, both of which are diagonals (*2*)
   (b) a quadrilateral with only 1 axis of symmetry, which is not a diagonal (*2*)
   (c) a figure with rotational symmetry of order 3. (*2*)
10.

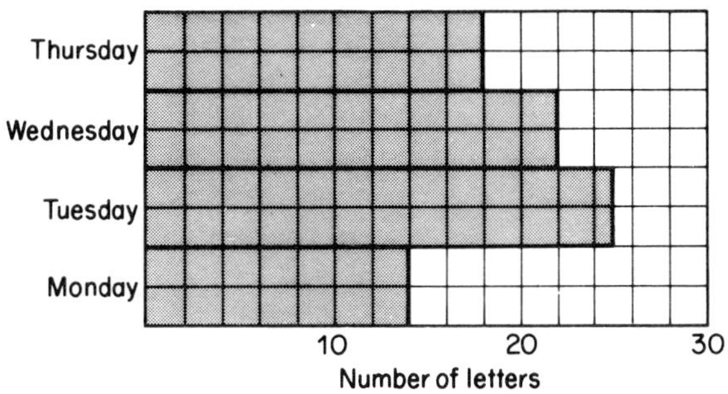

Figure 1

Figure 1 shows the number of letters sent by a small business on the first four days of a week.

(a) How many were sent on Wednesday? (*1*)
(b) On Thursday all letters were sent by 2nd class mail which cost 13 p for a letter. What was the cost of the postage on Thursday? (*3*)
(c) The mean (average) number of letters sent per day for the first *five* days was 20. How many letters were sent on Friday? (*4*)

Exercise 1.10

1. Calculate (**a**) $\frac{1}{2} + \frac{1}{3}$ of $\frac{1}{4}$ (*3*) (**b**) $\frac{2}{9}(4\frac{4}{5} + 1\frac{1}{2})$. (*4*)

2. Arrange $\frac{9}{25}$, 0.34, $\frac{7}{20}$ in ascending order of size. (Start with the smallest.) (*3*)

3. Helen earns £6500 a year as a designer. Her salary is about to be increased by 5%. What will she earn after the increase? (*4*)

4. Find the area of an isosceles triangle whose base is 7 cm and perpendicular height 6 cm. (*3*)

5. (**a**) Robin travels 5 km at 30 km/h. How long does the journey take? (*3*)
(**b**) George travels at 36 km/h for 40 minutes. How far does he travel? (*3*)

6. Factorise $a^2b - a^3b^2 + a^4b^3$. (*4*)

7. Solve (**a**) $\frac{x}{4} + 1 > 5$ (*3*) (**b**) $2(x - 3) = 4(x + 1)$. (*3*)

8. In a class of 20 children, 15 like apples. 10 like bananas, of whom 8 like apples also. How many like neither apples nor bananas? (A Venn diagram might help.) (*4*)

9.

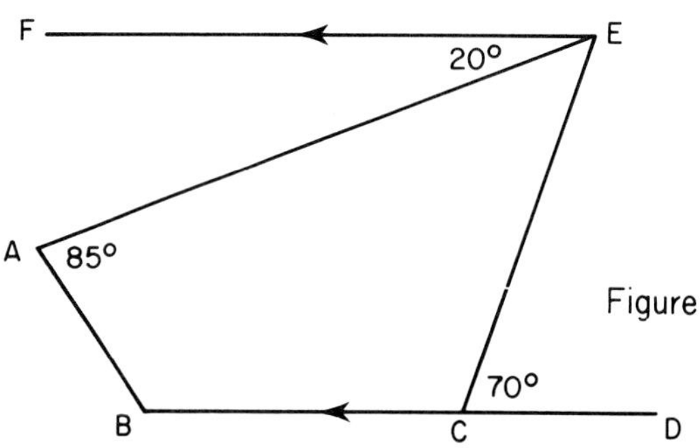

Figure 1

In Figure 1:
FE is parallel to BCD,
$\hat{FEA} = 20°$,
$\hat{EAB} = 85°$,
$\hat{ECD} = 70°$.

Calculate, giving reasons:

(**a**) $\hat{BCE}$ (*2*)

(**b**) $\hat{CEA}$ (*3*)

(**c**) $\hat{ABC}$. (*2*)

10. In Figure 2:

(**a**) Through what angle is triangle ABC rotated to give triangle XYC? (*2*)

(**b**) What is the equation of the mirror line in which triangle ABC is reflected to give triangle PQR? (*2*)

(**c**) What is the vector of the translation mapping triangle PQR onto triangle XYZ? (*2*)

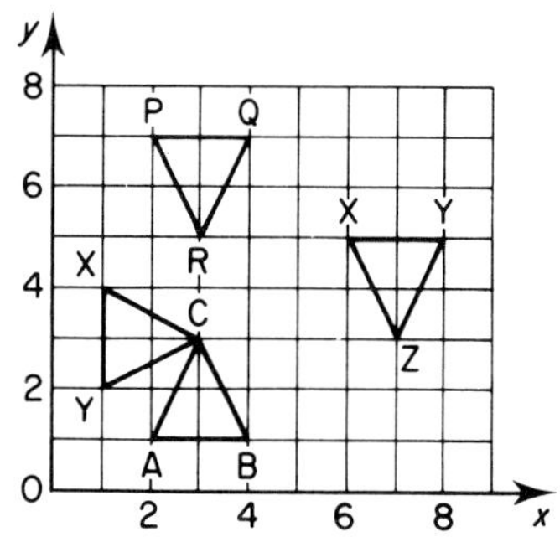

Figure 2

Exercise 1.11

1. Calculate $12.72 \div 5.3$ (*6*)
2. By using factors, find the square root of 784. (*4*)
3. Oliver and Paul get 14 marks each on a test. Jane, Karen and Lily get 17 marks each. What is the mean (average) mark of the five children? (*5*)
4. 140 cm of sticky tape is needed to wrap a parcel. How many parcels of the same size can be wrapped using a reel which has 12 m on it? (*5*)
5. **(a)** Write **(i)** 21_3 **(ii)** 42_8 **(iii)** 55_6 as denary (base 10) numbers. (*3*)
   (b) Add your answers to part (a) together in denary. (*1*)
   (c) Convert your answer to part (b) to base 4. (*2*)
6. Simplify **(a)** $(2a^2b)^3 \times 4a^3b$ (*3*) **(b)** $\frac{3}{a} - \frac{1}{2a} + \frac{2}{3a}$ (*3*)
7. If $c = 4$, $d = -2$ and $e = 0$, find the value of
   (a) $4c^2d^3e$ (*2*) **(b)** $(cd)^2 - cd^2$. (*4*)
8. Write a in terms of b, c and x, if $x = \frac{abc}{3}$. (*4*)
9. AB = 8 cm. X lies on AB such that BX = 3 cm. Using ruler and compasses only, construct a line at X which is perpendicular to AB. (*4*)
10. What is the size of an interior angle of a regular hexagon? (*4*)

Exercise 1.12

1. Calculate (**a**) $\frac{2}{5}$ of £1.60 (*3*) (**b**) $(3\frac{1}{3} - 2\frac{1}{2}) \div 1\frac{1}{4}$. (*4*)
2. Estimate $\frac{5.1 \times 4.2}{0.21}$. Give your answer correct to 1 significant figure. (*3*)
3. In a sale, a tailor lowers the price of his goods by 15%. What is the sale price of a jacket which originally cost £120? (*4*)
4. The perimeter of a rectangle is 56 cm and it is 8 cm wide. What is the area of the rectangle? (*4*)
5. It is 30 km from Mr Black's home to his office and a further 10 km to the warehouse. One morning he takes 40 minutes to drive to the office, where he spends half an hour talking to his boss, and then drives to the warehouse at 60 km/h. What is his average speed for the whole journey? (*6*)
6. Simplify $\frac{5x^2y + 10x^2y^2}{15x^2y}$. (*4*)
7. Solve (**a**) $3(x - 1) > 10$ (*3*) (**b**) $\frac{x}{4} - 3 = \frac{1}{2}$. (*3*)
8. Make two copies of the Venn diagram (Figure 1).
   (**a**) On the first copy, shade $X \cap Y' \cap Z$. (*2*)
   (**b**) On the second copy, shade $(X \cup Y) \cap Z$. (*3*)

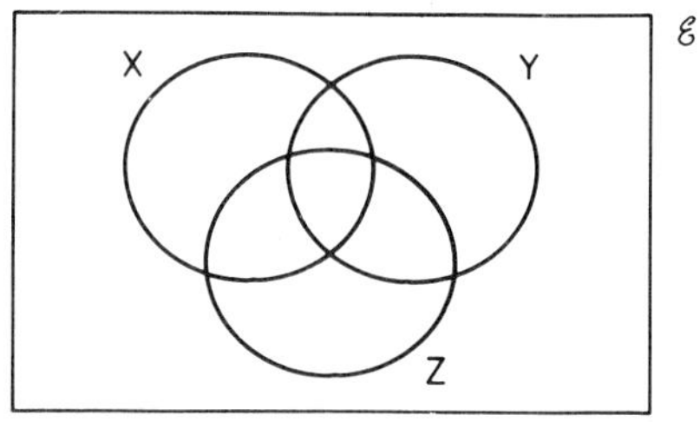

Figure 1

9. 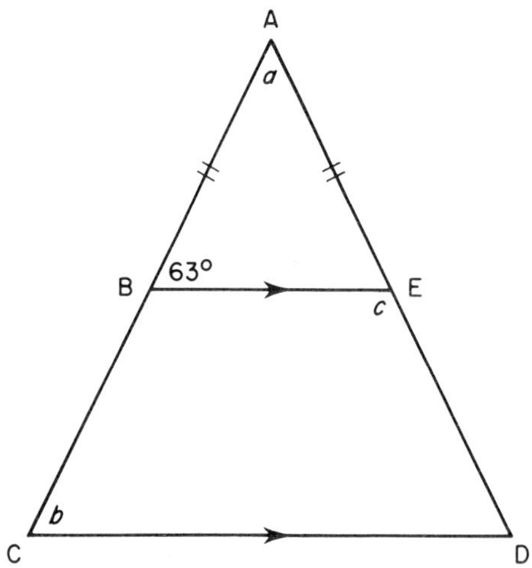

In Figure 2:
ABC and AED are straight lines.
BE is parallel to CD,
AB = AE,
$A\hat{B}E = 63°$.
Calculate, without giving reasons, the value of *a*, *b*, and *c*. (*6*)

Figure 2

10. (**a**) Mark *x* and *y* axes from 0 to 8. On these axes, mark A(3, 4), B(5, 6) and C(7, 4). ABCD is a square. What are the coordinates of D? (*2*)
    (**b**) CAEB is a parallelogram. What are the coordinates of E? (*2*)
    (**c**) F = (5, 0). What is the name given to the figure ABCF? (*1*)

2

Common Entrance Paper II: revision topics

Exercise 2.1 *Sets and Venn diagrams*

1. $\mathscr{E}$ = {20 children in form 3Z}
   B = {those who have a brother}
   S = {those who have a sister}
   2 have neither a brother nor a sister.
   9 have both a brother and a sister.
   Altogether 14 have a brother.

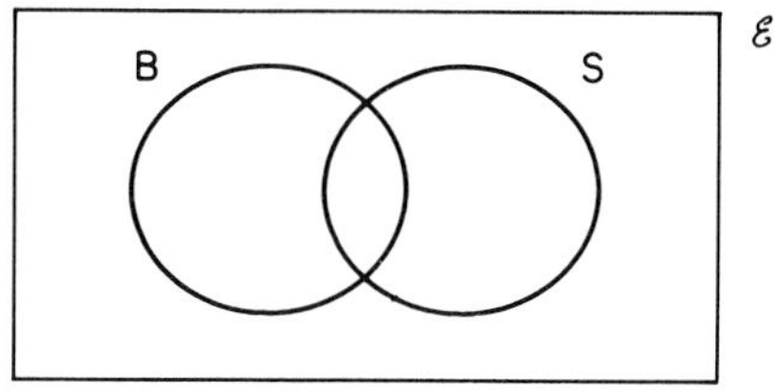

   (a) Copy the Venn diagram. Place the number of children in each part of the diagram.
   (b) How many children have a sister but not a brother?
   (c) How many members has B ∩ S′?

2. $\mathscr{E}$ = {C, O, M, P, U, T, E, R}
   A = {R, O, U, T}
   B = {C, U, T, E}
   (a) Copy the Venn diagram. Place the letters in the correct region.
   (b) What is *n*(A ∩ B)?
   (c) What are the members of (A ∪ B)′?

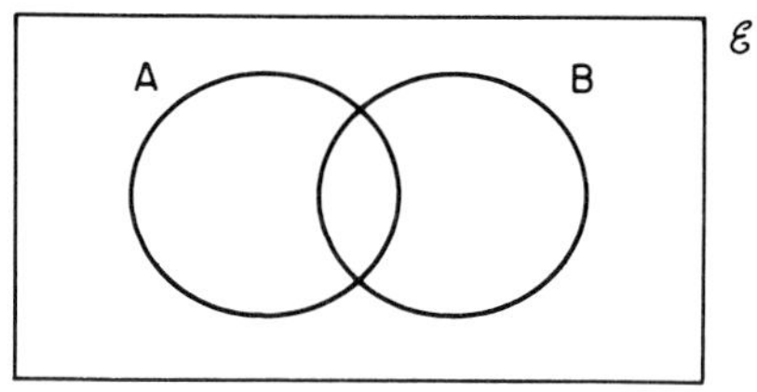

3. $\mathscr{E}$ = {books published by Arnold Edward}
   M = {Mathematics books}
   P = {paperback books}.
   State in plain English what you can say about the books in regions a, b, c and d.

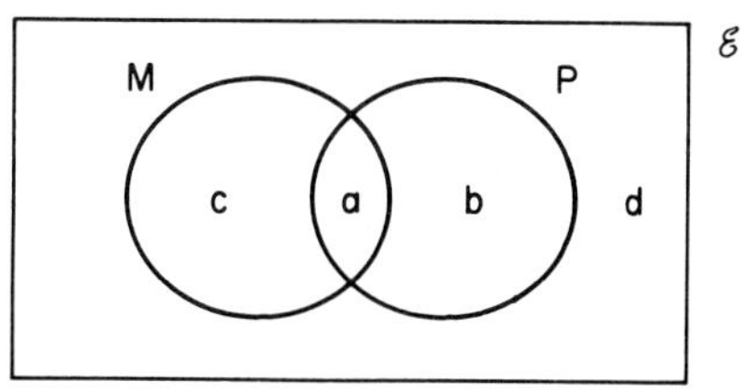

4. $\mathscr{E}$ = {whole numbers from 1 to 10 inclusive}
   A = {even numbers}
   B = {prime numbers} (1 is *not* prime.)
   (a) How many members has (A ∪ B)?
   (b) How many members has (A ∩ B)′?
   (c) What are the members of A′ ∩ B′?

5. $\mathscr{E}$ = {children at Holiday Hall}
   C = {those who play cricket}
   T = {those who play tennis}.
   There are 200 children at Holiday Hall, of whom 13 play neither game. $\frac{3}{5}$ of the children play both games and 180 play cricket. How many play tennis only?

6. $\mathscr{E}$ = {motor vehicles}
C = {cars}
D = {vehicles run on diesel fuel}
Draw a Venn diagram to represent this information.
(**a**) Anthony owns a diesel-driven car. Place a point labelled **a**, to represent Anthony, in the correct region of the diagram.
(**b**) Betty goes shopping on a petrol-driven scooter. Place a point labelled **b** in the correct region.
(**c**) Colin drives a diesel-powered bus. Place a point labelled **c** in the correct region.
(**d**) Denise fills up her car with 4-star petrol. Place a point labelled **d** in the correct region.
(**e**) S is the set of road rollers, which are all run on diesel. Draw and label S in the diagram.

7. On separate copies of the Venn diagram, shade in the following regions:

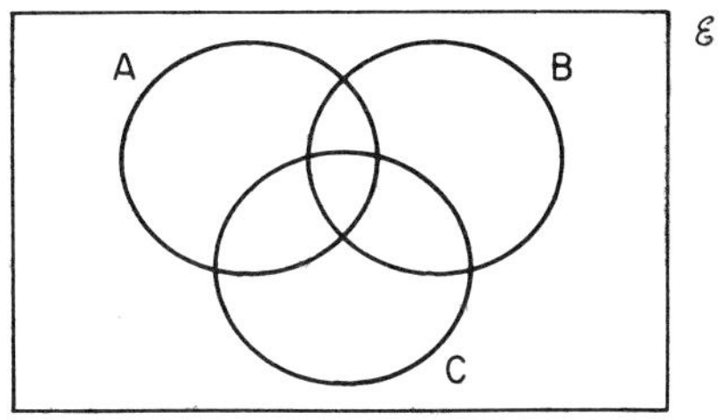

(**a**) $A \cap B$
(**b**) $A' \cap B \cap C$
(**c**) $A \cap B' \cap C'$
(**d**) $A \cup B$
(**e**) $(A \cup B \cup C)'$
(**f**) $(A \cap B) \cup C$
(**g**) $A \cup (B \cap C')$
(**h**) $A \cap (B \cup C)$
(**i**) $(A \cup B) \cap C$
(**j**) $A' \cap (B \cup C)$
(**k**) $(A \cap B)' \cap C$
(**l**) $(A \cup B)' \cap C$
(**m**) $(A \cap B) \cup C'$
(**n**) $(A \cap B) \cup (A \cap C)$
(**o**) $(A \cup B) \cap (A \cup C)$

8. Describe each of the shaded regions in set language:

(a)

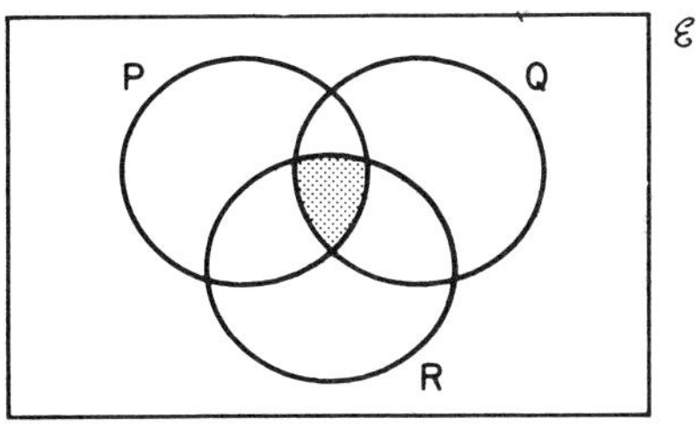

(b)

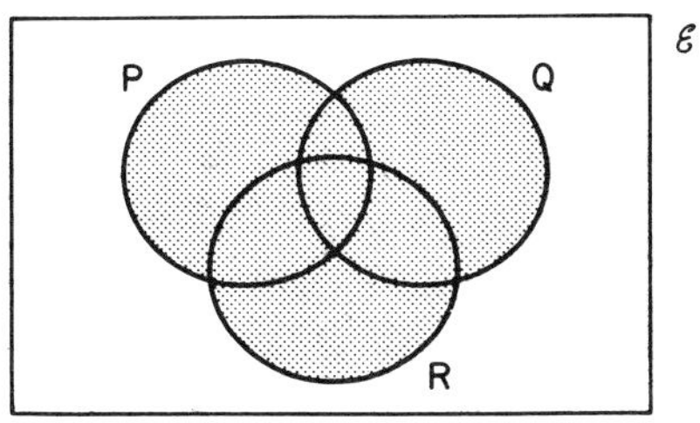

(c)

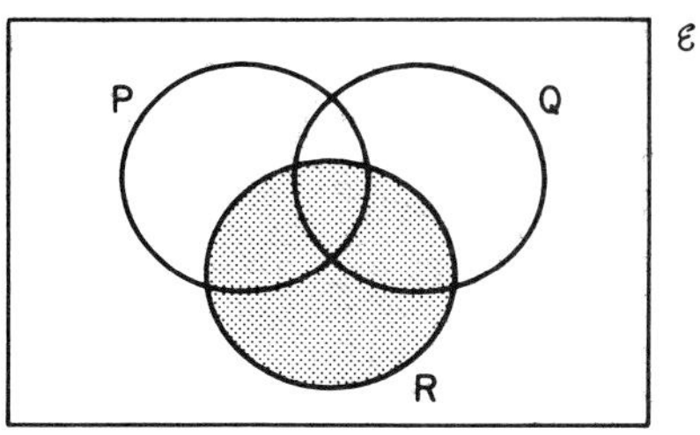

(d)

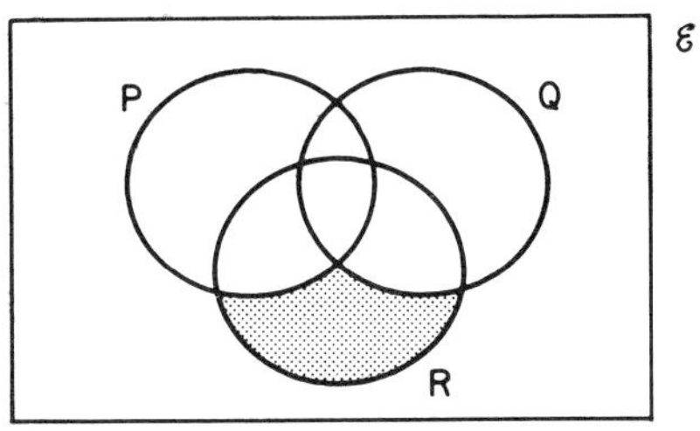

(e)

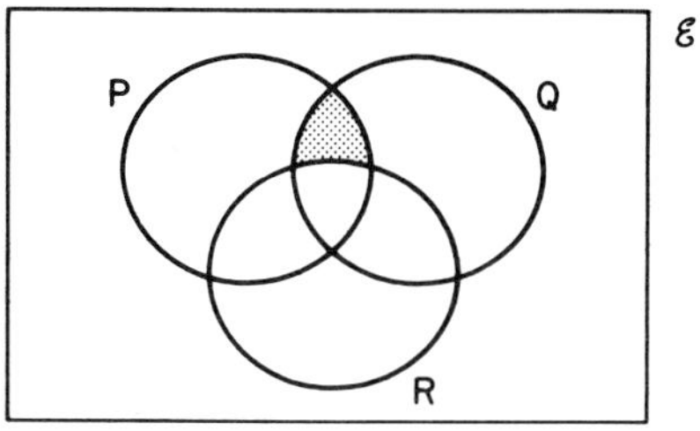

(f)

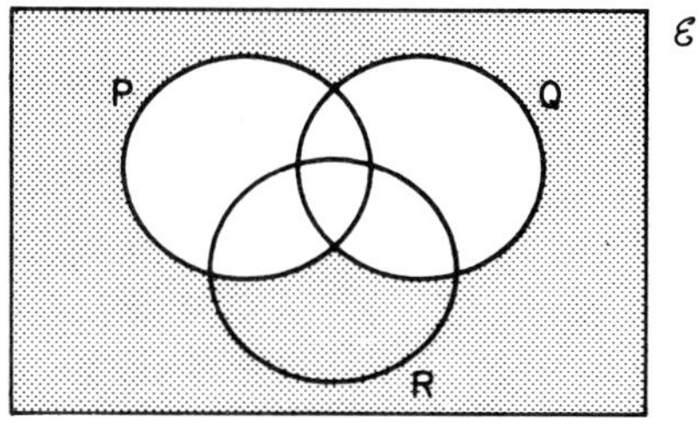

(g)

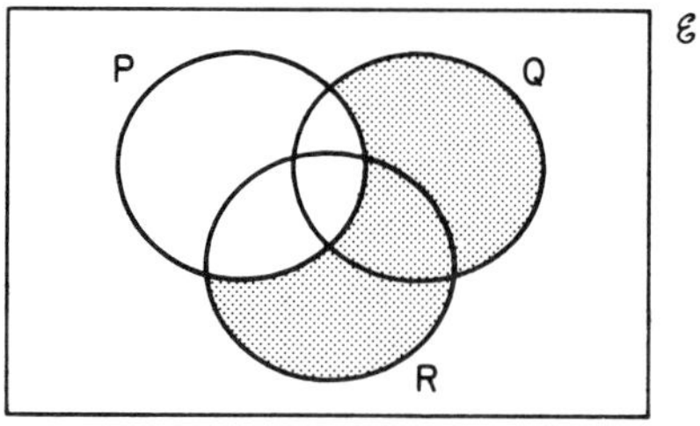

(h)

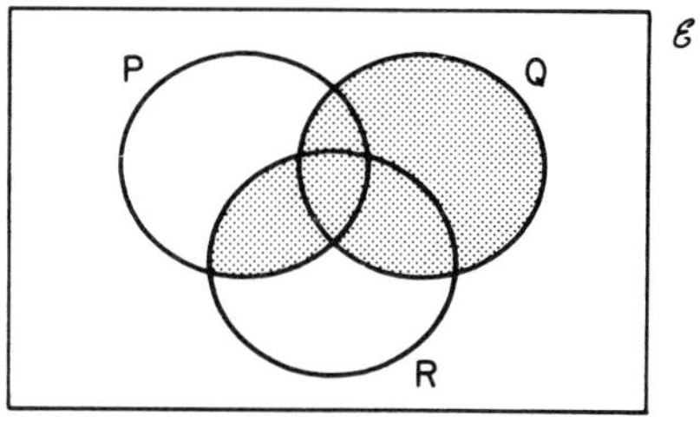

(i)

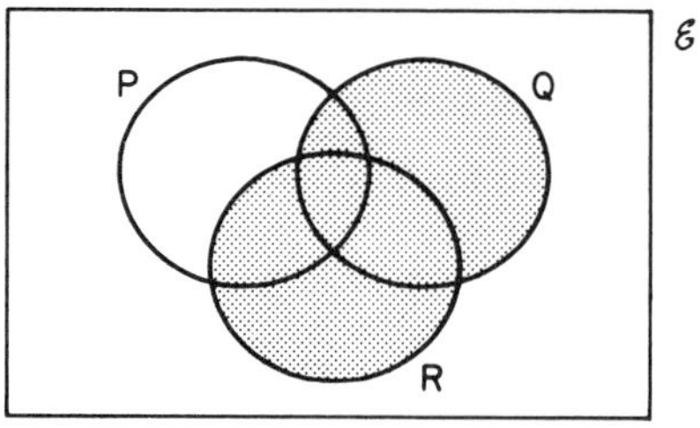

(j)

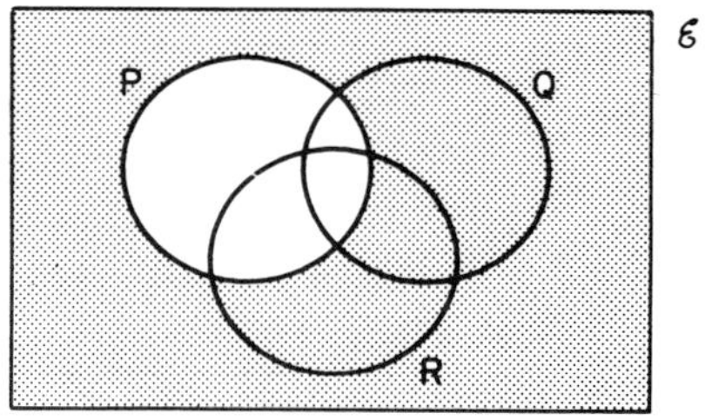

9. $\mathscr{E}$ = {positive integers $\leqslant$ 10}
A = {multiples of 2}
B = {multiples of 3}
C = {multiples of 4}

(**a**) What are the members of A, B and C?

(**b**) What is the relationship between (**i**) sets A and C
(**ii**) sets B and C?

(**c**) Draw a Venn diagram to illustrate $\mathscr{E}$, A, B and C, putting the numbers in the correct region.

10. On a Girl Guide picnic, every child took an apple, a banana or a pear to eat.
17 took an apple.
12 took a banana.
6 took a pear.
2 took all 3 fruits.
1 took a pear only.
2 took an apple and a banana but not a pear.
Nobody took a banana and a pear but not an apple.
Using the Venn diagram, find out how many Guides went on the picnic.

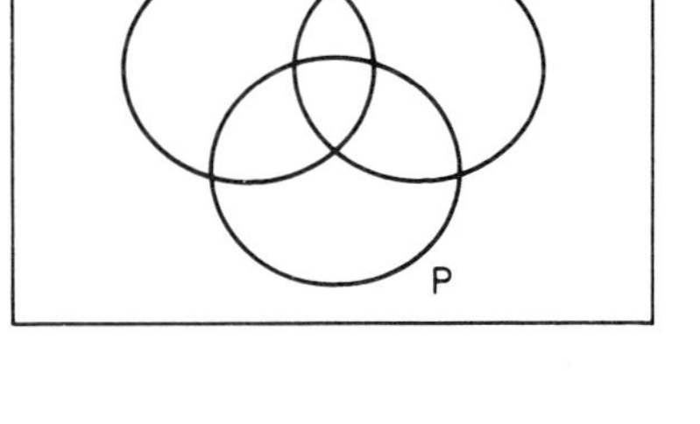

11. 135 travellers to the Channel Islands were asked whether they would be visiting Guernsey, Jersey or Sark.
58 said they would visit Guernsey.
83 said they would visit Jersey.
21 said they would visit Sark.
14 said they would visit Guernsey and Jersey, of whom 2 would go to Sark as well.
5 said they would visit Jersey and Sark but not Guernsey.
33 said they would visit Guernsey only.
(**a**) How many visited Jersey only?
(**b**) How many visited Guernsey and Sark but not Jersey?
(**c**) How many visited Sark only?
(**d**) How many did not visit any of these 3 islands?

12. A number of dog owners were asked which brand of food they gave their pets. All gave one or more of the following brands: Woof, Bonzo and Lap. No dog was given all three. 65 gave Woof; 56 gave Bonzo; 43 gave Lap; 50 gave Woof only; 6 gave Woof and Lap; 40 gave Bonzo only.
(**a**) How many dogs were given Woof and Bonzo?
(**b**) How many ate Lap only?
(**c**) How many dog owners were questioned?

13. In a form of 30, all do Physics, Chemistry or Biology. All who study Chemistry study Physics as well. 25 study Physics. 17 study Chemistry, of whom 8 learn Biology also. If 3 study Physics only, how many learn Biology?

14. A number of patients visited a hospital one week to have their ears, nose or throat examined. 20 had their ears looked at, 25 their nose and 30 their throat. 5 were checked for all three. Of those with good ears, 7 had a problem with both nose and throat. 8 altogether had their ears and nose looked at, the same number that had their ears only seen.
(**a**) How many had only their throat inspected?
(**b**) How many patients visited the hospital?

Exercise 2.2 *Perimeter, area and volume*

1. **(a)** A lawn measuring 20 m by 15 m is partly surrounded by a flower bed 3 m wide, as in the figure. Calculate the area of the flower bed.

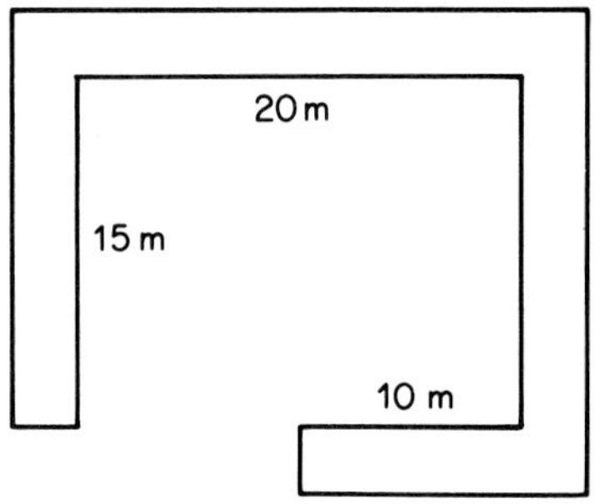

   (b) In 1 hour, a gardener can weed 18 m^2 of bed. How long will it take him to complete the job?

2. **(a)** The Browns' living-room measures 6 m by 4 m. Find the cost of carpeting the floor
   (i) with Axminster carpet at £15 per m^2
   (ii) with foam-backed carpet at £7 per m^2.
   (b) They decide to buy the Axminster, which comes in rolls of width 80 cm. To be most economical, should they lay the carpet along the room or across it?
   (c) Find the cost of a metre length of the roll of Axminster.
   (d) How many metres of the roll must they buy?

3. **(a)** Find **(i)** the perimeter, **(ii)** the area of a rectangular garden measuring 18 m by 12 m.
   (b) The garden is to be sown with grass and fenced. Sowing costs £2 per m^2. Find the cost of sowing.
   (c) Fencing costs £8 per metre. Find the cost of fencing and add this to your answer to part (b) to give the total cost.

4. **(a)** The figure shows a rectangular lawn surrounded by a path of width 2 m. Find the area of the path.
   (b) Write the ratio
   area of path: area of lawn
   in its lowest terms.

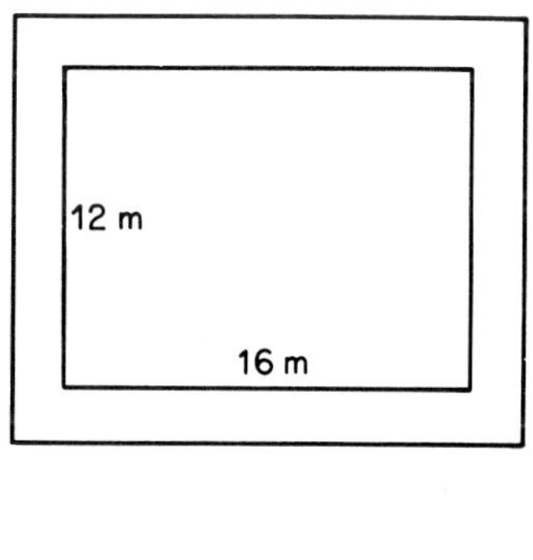

   (c) Also write the area of the path as a percentage of the total area of path and lawn.
   (d) It takes four times as long to apply weed-killer to 1 m^2 of path as it does to mow 1 m^2 of lawn. Harry starts mowing the lawn at 2 p.m. and finishes at 2.45 p.m. He immediately sets to work on the path with weed-killer. When will he finish?

5. (a) A rectangular lawn measures 44 m by 28 m. Find
(i) its perimeter, (ii) its area.
(b) Tidying the edge of the lawn takes a gardener 40 seconds per metre of edge. How long will it take him to tidy the whole edge?
(c) The lawn-mower mows strips 70 cm wide. Would you advise the gardener to mow parallel to the longer edges or parallel to the shorter edges?
(d) The mower moves at 1.1 m/s along each strip and takes 10 s to turn at each end. How long will it take the gardener to mow the lawn, mowing strips as in part (c)? Give your answer to the nearest minute.
(e) The gardener is paid £5.00 per hour. How much should he be paid altogether for tidying the edge and mowing the lawn? Round your answer up to the nearest £1 above.

6. (a) A living-room measures 6 m and 4.5 m and is 2 m high. Find the total area of walls and ceiling.
(b) In the walls, there are four doors each measuring 1.75 m by 0.8 m, a window measuring 1.25 m by 4 m and a fireplace measuring 2.4 m by 1 m. The remainder of the ceiling and walls are to be painted. Find the area to be painted.
(c) A tin of paint costing £7 will paint 14 m^2 of wall or ceiling. Find the cost of the paint required.

7. (a) The Haslams' bathroom measures 5 m by 3 m. The bath measures 2 m by 1 m and is positioned as in the figure. Find the area of the remainder of the floor.

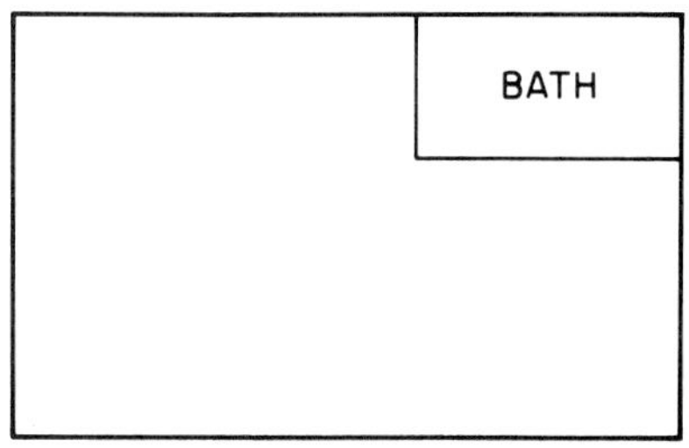

(b) The remainder of the floor is to be laid with linoleum. This comes in rolls of width 1 m, costing £2.50 per metre length. How much will Mr. Haslam have to pay for the linoleum?
(c) The room is 2.5 m high. Find the total area of the four walls.
(d) The bath is 80 cm high. Apart from the area behind the bath, the walls are to be tiled with square tiles of side 10 cm. Find how many tiles will be needed.
(e) Tiles cost 6p each. Find the total cost of tiles and linoleum.

8. (a) The figure shows three neighbouring fields. The angles at A, B, C and D are right angles. Calculate the area of each field.
(b) Write the ratio
area ABFE : area BCGF : area CDHG
in its lowest terms.
(c) It takes 1 h 30 min to plough the middle field. Find how long it should take to plough all three fields.

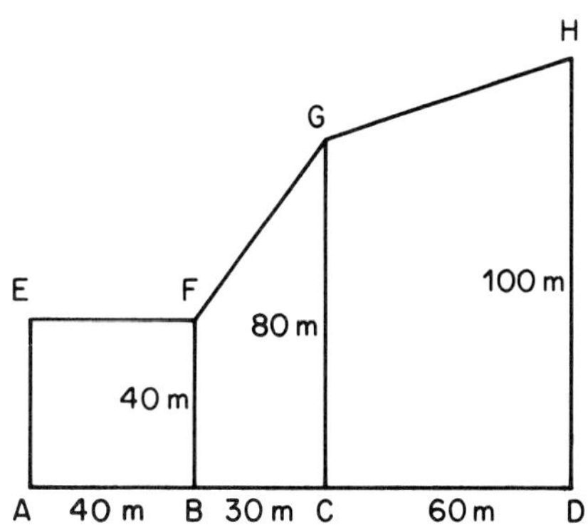

9. The figure shows a field ABCD, with AB = 100 m and right angles at B and C. It is divided by a fence DE where EA = x m.

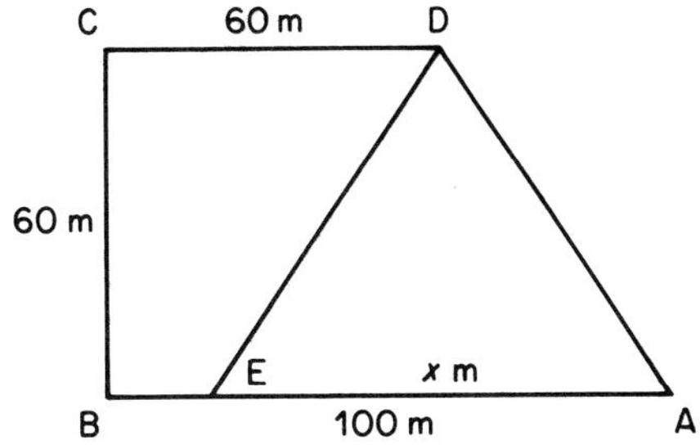

(a) Find the area of the whole field.

(b) Find the area of triangle DEA in terms of x.

(c) The fence DE divides the field into two equal parts. Find x.

10. **(a)** Little Haylocks School is laying a concrete cricket net. The concrete area must be 6 m long, 2.5 m wide and 30 cm deep. Find the volume of concrete needed, in m^3.

(b) It takes the concrete mixer 1 h 20 min to produce 1 m^3 of concrete. How long will it take to produce all the concrete required?

(c) To make 0.1 m^3 of concrete, mix a 50 kg bag of cement with 200 kg of gravel and 50 litres of water. Water is free, but a bag of cement costs £3.50 and gravel costs £30 per tonne. Find the total cost of making the concrete for the cricket net.

11. **(a)** Patrick Donohoe owns an old car, whose side-view may be approximately represented by the rectangular shape in the diagram. Both parts of the car are blocks of width 1.5 m. Find the volume of each block and so (ignoring the wheels) of the whole car.

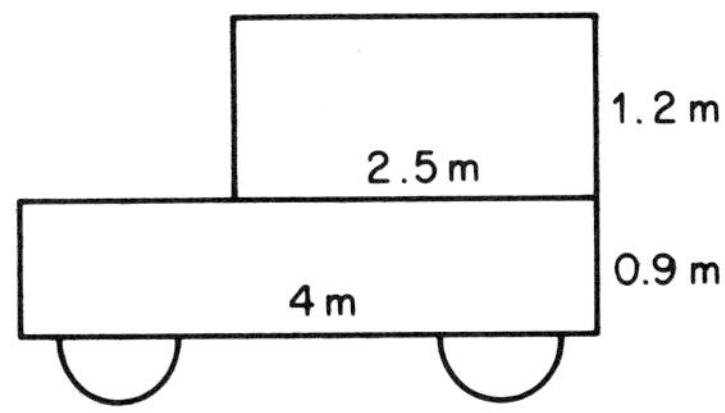

(b) Patrick keeps the car in a garage 5.5 m long, 4 m wide and 3 m high.

(i) Write the volume of the car as a percentage of the volume of the garage.

(ii) Find the volume of the space in the garage not occupied by the car.

(c) Patrick frequently leaves the engine of the car running with the garage doors closed. When this happens the garage fills with poisonous fumes at the rate of 1 m^3 every 20 seconds. How long will it be before the garage is completely full of fumes?

12. **(a)** A water tank has a rectangular base measuring 80 cm by 50 cm, and is 40 cm deep. How many litres of water will it hold when full?

(b) When the tank is empty, 50 litres of water is poured in. How deep will the water be?

(c) How much water must be added to raise the depth to 20 cm?

(d) Water is now scooped out using a scoop of capacity 75 cl. How many scoops will be needed to lower the depth to 5 cm?

13. **(a)** A car's fuel tank has a rectangular base measuring 50 cm by 40 cm, and is 30 cm deep. How many litres will it hold?

(b) When it is holding 10 litres of petrol, how deep is the petrol?

(c) When the tank is half full, the car is taken to a petrol station and is completely filled up in $1\frac{1}{4}$ minutes. Find the rate at which petrol is supplied, in cm^3/s.

14. (a) Fosbrooke School's oil storage tank has a rectangular base 2 m long and 1.5 m wide. Find the depth of oil when it contains 3300 litres.
(b) The tank is 2 m high. The school wishes to fill it up for the winter. How much extra oil should be ordered?
(c) How much will this oil cost, at 28p a litre?
(d) The tanker delivers the extra oil at the rate of 75 litres per minute. How long will it take to fill the tank?

15. (a) A garden swimming pool is 16 m long, 7.5 m wide and 2 m deep. Find how much water it will hold (i) in m^3, (ii) in litres.
(b) The pool can be filled at the rate of 250 litres per minute. Find how many hours it will take to fill the pool.
(c) 1 gram of chlorine must be added to every 1.5 litres of water. How many 50 kg loads of chlorine will be needed for the pool? Round your answer up to the nearest whole number above.

16. (a) A textbook measures 25 cm by 20 cm by 1 cm. Books are placed in boxes measuring 40 cm by 25 cm by 30 cm. How many books will fit into a box?
(b) The loading space of a van measures 2.5 m by 2 m by 1.8 m. Ignoring the thickness of the cardboard in the boxes, find how many boxes could fit into the van.
(c) Each book has mass of 150 g and sells for £4.50. Ignoring the mass and the value of the cardboard in the boxes, find (i) the total mass in tonnes of the load in (b) (ii) its value.
(d) In fact, the van is not permitted to carry a load of more than 1800 kg, and its contents are only insured for £55 000. Find
(i) how many boxes the van is permitted to carry
(ii) how many books it can carry
(iii) the mass, volume (in m^3) and value of its load.

17. (a) Find the total area of the bottom and sides of a swimming pool 15 m long, 8 m wide and 2 m deep.
(b) The pool is tiled using square tiles of side 20 cm. How many tiles will be needed?
(c) A paved area completely surrounds the pool to a width of 1.5 m. The paving tiles used are square and of side 25 cm. Find how many paving tiles are needed.
(d) Tiles of each kind come in lots of 50, at £20 per lot. How much will all the tiles cost?

18. The figure shows the side view of a house. The rectangle BCDE is 6.5 m long and 6 m high, and the triangle ABE is isosceles. The point A is 10 m above ground level CD.
(a) Find the height of the triangle ABE.
(b) Find the area ABCDE.
(c) The wall is built of bricks 20 cm long and 8 cm high. Calculate approximately how many bricks are used.

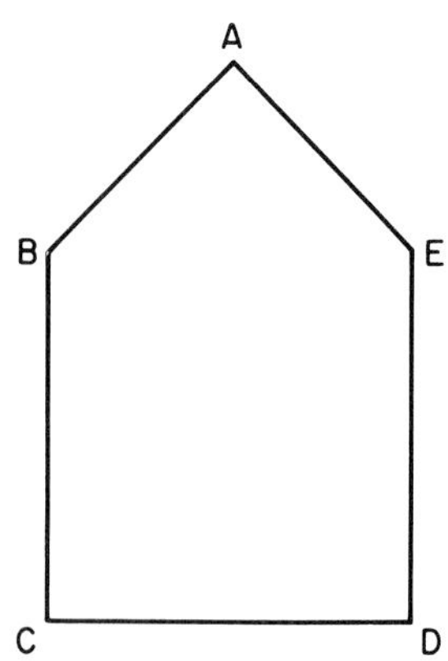

19. **(a)** Sunny, the hamster, is Rosemary's pet. The base of Sunny's cage measures 20 cm by 16 cm and it is 15 cm high. Pet books recommend, however, that hamster's cages should measure at least 30 cm by 20 cm by 20 cm. What percentage of the recommended minimum is the volume of Sunny's cage?

(b) Alarmed, Rosemary saves up for a new cage, and eventually buys one measuring 40 cm by 20 cm by 30 cm. The old cage had a single floor but this one has 3 floors, each 10 cm above the other. Write the ratio of Sunny's new living area to his old living area in its lowest terms.

(c) In the old cage, Sunny had a small nest in the shape of a cube of side 10 cm. Now, 10% of the volume of the new cage is devoted to his nest. Find the percentage increase in the volume of his nest.

20. **(a)** A Christmas cake is 26 cm long, 24 cm wide and 10 cm high. Each cm^3 of the cake has mass 0.8 g. Find **(i)** the volume of the cake, **(ii)** its mass, in kg.

(b) The top and sides of the cake are then covered in marzipan and icing to a total thickness of 2 cm. Find **(i)** the dimensions of the iced cake, **(ii)** its volume.

(c) Using (a) and (b), find the volume of the covering.

(d) All the vertical edges, and the perimeter of the top, are finished with fancy icing. If Mrs. Beston can do a 12 cm length of fancy icing in a minute, find how long she will take to complete the fancy icing.

Exercise 2.3 *Averages*

1. Sandra and Ruth were measuring how much further a spring stretched as they increased the load on it in steps of 100 grams each time. Their measurements were 3.4, 3.1, 3.3, 2.1, 3.3 and 3.4 cm.

(a) Sandra took all six readings and found the average. What was her answer?

(b) Ruth said that she thought one of the readings was wrong. Which do you think seems unreasonable?

(c) Ruth did not include the 'wrong' reading and found the average of the remaining five.

(d) What was the difference between their answers?

2. Sean and Stephen were competing for the 100 metres place in the athletics team. It was agreed that the runner with the lower average time for five races would win the place.

(a) Sean's times were 15.6, 14.9, 15.9, 15.0 and 14.6 seconds. What was his average time?

(b) Stephen's times were 16.3, 14.2, 13.5, 16.4 and 14.8 seconds. What was his average time?

(c) Who won the place?

(d) If these times were taken in the same races, on how many occasions did Sean beat Stephen?

(e) If they had agreed that each boy would discard his best time, who would have won the place?

3. The scale measuring the depth of a river above and below normal was marked off as in the diagram on the right. The readings for a certain week were:

   −7.5, −4.8, −0.6, +1.2, +0.8, +0.3, +0.1

   (a) What was the average reading for the week?

   (b) Would you say that the weather became more or less wet towards the end of the week? Give reasons.

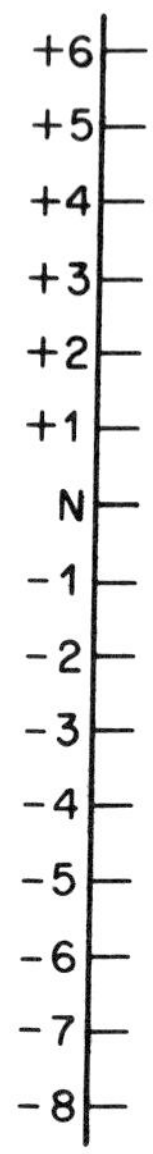

4. Clive scores 411 runs in 15 completed innings.

   (a) What is his average number of runs per innings?

   (b) In his next innings, he scores enough runs to bring his average up to 28.5 before being bowled. How many runs did he score in this innings?

   (c) What would his average have been if he had not been out in this latest innings? [The batting average is (total number of runs) ÷ (number of times out).]

5. The average height of Mr and Mrs Wadlow and their son Robert was 201 cm in 1939.

   (a) If Mr and Mrs Wadlow had an average height of 171.5 cm, what was the height of their son Robert, the tallest recorded living person?

   (b) If his father and mother had masses of 72 kg and 50 kg respectively, and the average mass for the whole family was 114.9 kg, what was Robert's mass?

6. The combined marks for Maths I and Maths II for six boys in an exam were 27, 153, 120, 56, 41, 95.

   (a) Find the average total per boy.

   (b) Find the average mark per paper.

   (c) If both papers were out of 80, what was the average percentage per paper?

7. The number of people per day who paid £1.50 to enter a museum were 536, 441, 272, 213 and 363. The museum was shut on Sundays and Thursdays.

   (a) What was the average number attending per day?

   (b) How much money was taken on each day?

   (c) What was the average amount of money taken per day?

   (d) Can you suggest another way of arriving at the average amount of money taken per day?

8. Tamsin Jones has to get up by 8.15 a.m. if she is to catch the school bus. Her getting-up times for a week were 10.20, 7.43, 8.10, 7.40, 7.50, 8.00 and 9.40 a.m.
   (a) Find the average time that Tamsin got up.
   (b) On average, would you expect her to catch the bus or not?
   (c) Giving reasons, say which day of the week you think that they began recording her getting-up times.
   (d) Why do you consider that your result for the average time might be misleading?
   (e) Are there any times that you think should be left out, if your answer to (b) is to have any meaning?

9. Ian has taken 35 wickets at an average of 12.4 runs per wicket.
   (a) How many runs have been scored off his bowling so far?

   In his next match he achieves another 3 wickets at an average of 20.0 runs per wicket.
   (b) How many runs have been scored off his bowling now?
   (c) What is his new average per wicket?
   (d) 31 runs are scored off his bowling in the next match. How many wickets would he have to take to bring his final average to 12.5 runs per wicket?

10. Belinda went on a cycling tour for seven days. On average, she cycled 41.2 km per day.
   (a) What was the total distance that she cycled in the week?
   (b) On the first two days she cycled 52.2 km per day, on average. What was the total distance for these two days?
   (c) What was the distance she covered in the last five days?
   (d) What was the average distance she travelled per day for the remaining five days?

11. A man travels for four days. On the first day he travels 64 km at an average speed of 24 km/h. On the second day he travels at an average speed of 30 km/h for 3 hours 20 minutes. On the third day he travels 88 km in $2\frac{2}{5}$ hours.
   (a) How long does his journey take on the first day?
   (b) How far does he travel on the second day?
   (c) How far does he travel altogether in these three days?
   (d) For how long is he travelling?
   (e) What is his average speed while travelling, for the whole three-day journey?
   (f) If on the fourth day he makes his total distance up to 300 km, in what length of time must he complete his journey to keep the same average speed as before?

12. Christopher, Adam and Jennifer spend £5.60, on average, on presents for their parents at Christmas. Christopher bought some soap for his mother for £3.20 and a scarf for his father for £3.90. Robert spent £4.10 on a pair of gloves for his mother and £2.60 on a tie for his father.
   (a) Without calculating the exact value, show whether Jennifer spent more or less than the average.
   (b) Jennifer bought a plant as a combined present for her parents. How much did it cost?
   (c) If she had spent 63 pence more on the plant, what would the average amount spent per person now come to?

13. Mr Potter budgeted to spend £6.50 per week on petrol. In the first month he bought 3.3, 4.0, 2.5 and 3.8 gallons per week.

(**a**) Calculate the average number of gallons per week for this month.

(**b**) If the petrol cost £1.90 per gallon, did he keep to his budget?

(**c**) If the cost increased to £1.95 per gallon and Mr Potter refused to increase his budget, how many gallons a week could he now buy, on average?

(**d**) If he bought 3.5, 2.8, 3.0 gallons in the first three weeks at this new price, what was the most that he could buy in the fourth week, if he was to keep to his budget? Give your answer to the nearest tenth of a gallon.

14. A class of 12 pupils has an average age of 12 years 9 months. At the beginning of the new term the class is joined by three more pupils who have an average age of 12 years 4 months.

(**a**) What is the total age of the class of 12 pupils?

(**b**) What is the total age of the three new pupils?

(**c**) What is the total age of all 15 pupils?

(**d**) What is the new average age of the whole class?

15. A group of eight people have masses of 43.2 kg, 56.1 kg, 46.2 kg, 42.7 kg, 43.0 kg, 52.1 kg, 50.6 kg and 45.3 kg.

(**a**) What is the total mass of the group?

(**b**) Find the average mass per person.

(**c**) The party needs to use a lift with a load restriction of 190 kg maximum. Can it transport everyone in just two groups? Give calculations to justify your answer.

16. Paula's father manages a supermarket. The number of miles travelled per day for the three delivery vans is entered in a form, as shown below. Unfortunately, Paula's father found at the end of one week that some entries had not been filled in.

	Mon	**Tues**	**Wed**	**Thur**	**Fri**	**Average per day**	**No. of miles/litre**
Van A	21	29	23	14	46		2.5
Van B	32	17	39	28		29.6	2.0
Van C	25		28	32	15	24.8	2.6
Average per van	26	$23\frac{1}{3}$	30		31		

(**a**) What was the daily average for Van A?

(**b**) How many miles did Van B do on Friday?

(**c**) What was the average distance travelled on Thursday?

(**d**) What was the distance travelled by Van C on Tuesday?

(**e**) On average, which van, A or C, uses more petrol during the week?

17. Mr Crankshaw, the form-master, dropped his mark sheet for the eight boys taking an examination this term. The rain smudged some of the figures but Edwinson spotted that all the missing results could be filled in from the remaining figures. Can you do it?

Name	English	Geog	Hist	Maths	French	Average mark
Armingly	72	64	52	68	76	
Baines	46	52	64		51	48.0
Crawson		58	66	82	60	65.4
Dibley	54	46	30	48	51	45.8
Edwinson	84	76	69	72	81	76.4
Fairsonly	68	70	64	26	58	
Grossfield	26	45	37	55		42.8
Heskworth	37	53	50	46	28	42.8
Average mark per subject	56		54	53		

(a) Calculate the average mark for Armingly.
(b) Calculate the Maths mark for Baines.
(c) Calculate the English mark for Crawson.
(d) Calculate the average mark for Geography.
(e) Giving reasons, show how you would find an approximation for Fairsonly's average mark.
(f) Calculate the correct average mark for Fairsonly.
(g) Calculate Grossfield's mark for French.
(h) Calculate the average mark for French.
(i) Taking the average mark for History, make two columns to show how much above or below the average each boy's History mark is. Add up the total in each column. What do you notice?
(j) Now try this idea for the average mark for Baines.

Exercise 2.4 *Speed*

1. Find how far these people travel.

(**a**) Alan cycles for 3 hours at 12 km/h and then walks for 2 hours at 4 km/h.

(**b**) Barbara jogs for 1 hour 45 minutes at 8 km/h.

(**c**) Charles runs for $1\frac{1}{2}$ minutes at 7.5 m/s.

(**d**) Mr Dewhurst drives for $1\frac{1}{2}$ hours across London at 32 km/h and then for a further 2 h 20 min on the motorway at 96 km/h.

(**e**) Elizabeth walks down the High Street for 5 minutes at 1.5 m/s and then runs for a bus for 20 s at 6 m/s.

2. Find how long these journeys take.

(**a**) Anabel canoes 6 km up the river at 9 km/h.

(**b**) Bernard runs to the station, 1.5 km away, at 4 m/s.

(**c**) Mrs Cripps goes round for coffee to Mrs Calver, 2.4 km away, walking at 4.5 km/h.

(**d**) Debbie cycles 3 km to Diana's house at 15 km/h. They then cycle together for 20 km at 12 km/h.

(**e**) Mr Edwards drives to the station, 4 km from his home, at 30 km/h. He catches a train, which travels for 36 km at 80 km/h, and then walks 2.4 km to his office at 1.6 m/s.

3. At what speed to these people travel?

(**a**) An athlete runs the 400 m in 50 s.

(**b**) A racing driver travels 3840 km in 24 hours.

(**c**) A woman drives to her job in the market town, 24 km away, in 36 minutes.

(**d**) A yachtsman sails 65 km across the Channel in $12\frac{1}{2}$ hours.

(**e**) A pilot flies 819 km in 1 h 18 min.

4. The journey from Combridge to Munchester is partly on trunk roads and partly on a motorway.

(**a**) Mike Hawkins can average 72 km/h on the 108 km of trunk roads. How long will he spend on the trunk roads?

(**b**) The whole journey takes him 4 hours. He can average 100 km/h on the motorway. How much of the journey is on the motorway?

(**c**) Find the total length of the journey.

(**d**) Calculate Mike's average speed for the whole journey.

5. A tramp steamer can sail at 24 km/h when the weather is calm but can only manage 9 km/h in rough seas.

(**a**) It sets out from Port Musgrove at 10 a.m. one day and heads towards Nettlesham, 150 km away. At 2.30 p.m. it hits rough weather.

(**i**) How far has it gone when it hits the rough weather?

(**ii**) Find when it reaches Nettlesham.

(**b**) The following morning it sets out for Harmouth at 7 a.m.

(**i**) The sea is rough for the first 33 km. At what time does it meet calm weather?

(**ii**) After this, the weather remains calm. It reaches Harmouth at 5 p.m. How far is Harmouth from Nettlesham?

6. (**a**) A train leaves Meeston station at 0930 and is timetabled to travel to Lemon Street at an average speed of 120 km/h. It is 400 km from Meeston to Lemon Street. Find when the train is timetabled to arrive at Lemon Street.

(**b**) It travels at 120 km/h until it reaches Teame at 1145. How far is it from Meeston to Teame?

(**c**) Because of modernisation at Teame, the train has to travel at only 40 km/h for the next 15 minutes. How far will it be from Lemon Street after this delay?

(d) At what speed must it now travel if it is to reach Lemon Street on time?

7. A sponsored walk takes place over a 6 km course. Participants are expected to complete five laps of the course.

(a) Barry Hobble does the first four laps at 4 km/h and then speeds up to 5 km/h for the final lap. How long does he take to complete the course?

(b) Mara Thon runs the first three laps at 2.5 m/s. After that she strains an ankle and has to walk at 40 m/min. How long does she take to complete the course from start to finish?

(c) How far has Barry still got to walk when Mara finishes?

8. (a) It is 9100 km by air from London to Hong Kong. One airliner completes the flight non-stop at 650 km/h. How long does the flight take?

(b) Another airliner sets off on the same journey at 700 km/h. After flying for $7\frac{1}{2}$ hours, it stops to refuel. How far is it then from Hong Kong?

(c) After half an hour it resumes its journey, which takes a further 7 hours. How fast does it fly on this part of the journey?

(d) Calculate the average speed of the second airliner for its whole journey, giving your answer in km/h correct to 2 significant figures.

9. (a) Over level ground, Caesar's troops could cover 40.5 km in a $7\frac{1}{2}$-hour day's march. At what speed could they march over level ground?

(b) Over hilly country they could only march at 3.5 km/h but they would march for 8 hours in a day. How far could they march in a day over hilly country?

(c) The first 10.5 km of the journey from Otega to Latvium is hilly; after that it is level. The whole journey takes Caesar's troops 8 hours. Find the distance from Otega to Latvium.

10. Two European champion balloonists, Igo Reitup and Ivan Heier, are racing each other to an altitude of 30 000 m.

(a) Igo climbs at a steady rate of 2 m/s. Find how long he will take. Write your answer in hours and minutes.

(b) How many km is 30 000 m?

(c) Ivan ascends at 6 km/h for 2 hours but then runs into unfavourable air currents and gains no height for the next half hour.

(i) How many metres has he still got to climb?

(ii) How long has he still got before Igo reaches 30 000 m?

(d) Find, in metres per second, the minimum speed at which Ivan must now climb if he is to prevent Igo from winning.

11. Anne and Clara agree to meet at Beeston at 1200.

(a) Anne lives 98 km from Beeston and can drive at an average speed of 56 km/h. At what time should she leave home?

(b) Clara leaves her home at 1000. She drives at an average speed of 72 km/h and arrives at Beeston at 1045. How far from Beeston does she live?

(c) Having time to spare, Clara goes for a drive at the same average speed as before. How far away from Beeston can she go if she is still to get back on time?

12. (a) A power boat covers the 90 km from Eastbourne to Portsmouth at 40 km/h. Find how long the journey takes.
(b) After $1\frac{1}{2}$ hours rest it starts back for Eastbourne. However, it has engine trouble, and takes $1\frac{1}{4}$ hours to cover the first 15 km. It then completes the remainder of the journey at 30 km/h. Find how long the return journey takes.
(c) Calculate the boat's average speed for the entire round trip, including the rest.

13. Anna and her sister Kate line their snails up for a race. The winner is the snail that is ahead after 2 minutes.
(a) Anna's snail, Conchorde, gets off to a good start, covering 7.5 cm in the first 30 seconds. What is Conchorde's speed, in cm/s?
(b) Unfortunately, Conchorde moves 1 cm backwards in the next 30 s. Anna then gets it moving in the right direction at 0.3 cm/s until the end of the race. How far from the start will it be at the end of the race?
(c) Kate's snail, Caravolle, does not move at all for 20 s but then heads forward at a steady 0.28 cm/s. Which snail wins the race, and by how much?

14. Fred drives a train from Wilton to Barville, a distance of 100 km. He leaves Wilton at 1125 and is due at Barville at 1240.
(a) Fred travels at a constant speed and arrives at Barville exactly on time. At what speed does he travel?
(b) S is a signal-box between Wilton and Barville, 40 km from Wilton. At what time does Fred pass the signal-box?
(c) Len drives a train from Barville to Wilton at a constant speed of 90 km/h. At what time does he leave Barville if he passes Fred at S?

15. (a) When racing, a rowing eight can cover the first 3.6 km in 10 minutes. Find its speed in m/s.
(b) The total length of the race is 6480 m. If the eight continued at the same speed, how long would it take to complete the rest of the course?
(c) In fact it takes another 10 minutes to finish the race. At what fraction of its earlier speed is it travelling?
(d) Find its average speed over the whole race, in m/s.

16. (a) To beat the record in the Barville annual cycle race, you would have to maintain an average speed of 36 km/h for 2 h 40 min. How long is the course?
(b) Michael is determined to beat the record. He discovers he can keep up an average speed of 45 km/h on level ground. In the race, there is 60 km of level ground. How long will this take him?
(c) He practises on the hills and finds he can do 50 km in 1 hour 40 minutes. Find his average speed on hills.
(d) How many km of hills are there on the course?
(e) How long will it take Michael to cover this distance at the same average speed as in (c)?
(f) Find whether Michael succeeds in beating the record, and if so by how much.

17. Two yachts, Barbuda and Tahiti, compete in a yachting race over a triangular course.

(**a**) The first leg is a downwind run of 12 km. Here, Barbuda has the advantage and completes the leg in 1 h 20 min. Find her average speed.

(**b**) Tahiti can only manage 8 km/h on this leg. Find how much longer she takes over the first leg than Barbuda.

(**c**) On the second leg, the ships are well matched. They each complete it in 1 h 20 min at a constant speed of 12 km/h. Find the length of this leg.

(**d**) The final leg is an upwind tack of 20 km. Tahiti manages an average speed of 6 km/h on this leg. Find how long she takes over this leg.

(**e**) Barbuda takes a total of 6 h 20 min over the whole course. Who wins the race and by how many minutes?

18. (**a**) It is 400 km from Manchester to Exeter on the motorway. The Harrises set out at 0800 and average 90 km/h for the first 225 km, when they stop for coffee. At what time do they stop for coffee?

(**b**) Half an hour later, they resume their journey and reach Exeter at 1245. Find their average speed for this part of the journey.

(**c**) The car uses 8 litres of fuel per 100 km at 90 km/h and 10 litres per 100 km at the speed in (b). Find how much fuel it uses over the whole journey.

(**d**) Find also the cost of this fuel, at 40 pence per litre.

19. A business woman has the choice of two methods of travel from London to Edinburgh.

(**a**) One is to take the train, which covers the 480 km in 5 hours. Find the speed of the train, assuming that it is constant.

(**b**) Otherwise she can fly. The plane covers 490 km at an average speed of 350 km/h. Find how long the flight takes.

(**c**) Unfortunately, flying also involves travelling a total of 60 km in taxis at an average speed of 36 km/h, and a further delay of 20 minutes at each airport. Find the total time spent in taxis and at airports.

(**d**) How much quicker is flying, overall?

20. (**a**) In a cycle slow-race over 120 m, Bertie does the first 30 m at 0.4 m/s. How long does he take over this?

(**b**) He then cycles at 0.3 m/s for the next $2\frac{1}{2}$ minutes. How far does he go?

(**c**) He finishes off the race in a further 3 minutes. What is his speed over this part of the course?

(**d**) His brother Edward travels at a constant speed of 0.3 m/s for the whole race. Who wins and by how many seconds?

Equations

1. There are 24 000 litres of water in one tank and 5000 litres in another. Equal quantities of water are poured into each tank until the first contains twice as much as the second.
   - **(a)** If y litres of water are added to each, how many litres will there be in the first tank, in terms of y?
   - **(b)** How many litres of water will there be in the second tank, in terms of y?
   - **(c)** Form an equation and use it to find the value of y.
   - **(d)** How many litres of water will there be in each tank when the first tank is twice as full as the second?
2. A man is twice as old as his son. 15 years ago he was three times as old.
   - **(a)** Taking the son's age as x years, how old is the father now, in terms of x?
   - **(b)** How old was the son 15 years ago, in terms of x?
   - **(c)** How old was the father 15 years ago, in terms of x?
   - **(d)** Form an equation for their ages 15 years ago and solve it to find the value of x.
   - **(e).** What is the father's age now?
3. The number killed, wounded and taken prisoner in a certain battle amounted to 221. The number of prisoners was 7 times the number killed, and 11 more were wounded than taken prisoner.
   - **(a)** Letting the number killed be x, how many were taken prisoner, in terms of x?
   - **(b)** How many were wounded, in terms of x?
   - **(c)** Form an equation and solve it to find the value of x.
   - **(d)** How many were wounded in the battle?
4. Anne, Barbara and Christine have £114 between them. Anne's share is double that of Barbara and Christine's is one sixth of Barbara's.
   - **(a)** If Barbara has £$6x$ (why $6x$?), then what will the others have, in terms of x?
   - **(b)** Form an equation for the total amount and use it to find the value of x.
   - **(c)** How much money does each have?
   - **(d)** Check that the total amount is £114.
5. The price of coal is reduced by 30%. The new price is £3.57 a bag.
   - **(a)** Taking the original price to be $100x$ pence (why $100x$?), what is the price after it has been reduced by 30%, in terms of x?
   - **(b)** Form an equation for the new price and solve it to find x.
   - **(c)** What was the price per bag originally?
6. A man invests £100 and receives £3 per year interest on it. His son invests £80 but receives £5 per year interest.
   - **(a)** After y years, how much will the father have, in terms of y?
   - **(b)** After y years, how much will the son have, in terms of y?
   - **(c)** Given that, after y years, these amounts are equal, form an equation and solve it to find the value of y.
7. Daffodil bulbs cost 4p each and tulip bulbs cost 30p each. Mr Thomas buys 560 bulbs altogether at a total cost of £64.
   - **(a)** If he bought d daffodil bulbs, how many tulips did he buy, in terms of d?
   - **(b)** How much money did he pay for the daffodils, in terms of d?

(c) How much money did he pay for the tulips, in terms of d?
(d) Form an equation for the total cost and use it to find how many daffodils he bought.
(e) How much did he spend on (i) daffodils, (ii) tulips?

8. A boy is out in each of his first t innings. His average is 18 runs per innings. In his next innings he is bowled for 42 and increases his average by 2.
(a) How many runs did he score in his first t innings?
(b) How many runs did he score in $(t + 1)$ innings?
(c) Find another way of expressing his total after $(t + 1)$ innings.
(d) Form an equation for his total after $(t + 1)$ innings.
(e) Solve it to find the value of t.
(f) How many runs did he actually score in $(t + 1)$ innings?

9. June is 8 years older than her sister Betty, and 6 years older than her brother Paul. June is t years old. Twice June's age added to the total age of Paul and Betty is ten times Betty's age.
(a) What is Betty's age, in terms of t?
(b) What is Paul's age, in terms of t?
(c) What is the sum of the ages of Paul and Betty, in terms of t?
(d) Form an equation in t.
(e) Solve it to find June's age.

10. A full return costs $1\frac{3}{4}$ times as much as a cheap day return for the same distance. The difference in cost is £4.20.
(a) If a cheap day return costs x pence, what is the cost of a full return, in terms of x?
(b) What is the difference in cost, in terms of x?
(c) Form an equation for the difference in cost.
(d) Solve this equation to find the value of x.
(e) What is the cost of a full return?

11. Alice and Simon go on a family holiday with their parents. Alice has £14 more than Simon, to start with. At the end of a week, Alice has spent one fifth of her money and Simon a quarter of his. They have £28.25 left between them.
(a) How much did Alice have left, if she started with £a?
(b) How much did Simon have to start with, in terms of a?
(c) How much did Simon have left, in terms of a?
(d) Form an equation for the amount remaining and use it to find the value of a.
(e) How much did each start with?

12. A man buys a number of bicycles at £52 each. He sells two thirds of them at £73 each and three at £71 each. He then sells off the remainder at £60 each. He makes an overall profit of £283.
(a) If the man bought b bicycles at first, how much did he pay for them altogether?
(b) How much did he receive for those he sold at £73, in terms of b?
(c) How much did he receive for those he sold at £71?
(d) How many were left, in terms of b?
(e) How much did he receive from the sale of the £60 bicycles?
(f) Form an equation for the total amount spent and received – don't forget the profit.
(g) Use this equation to find how many he sold at £73.

13. From Marton to Carbury is 19 km. David starts to walk from Marton to Carbury at $3\frac{1}{3}$ km/h. Susan starts half an hour later from Carbury to cycle to Marton at 10 km/h.

(a) If David has travelled for t hours when they meet, how far will he have travelled, in terms of t?

(b) For how long will Susan have been travelling when they meet, in terms of t?

(c) How far will she have travelled in this time, in terms of t?

(d) Form an equation for the sum of the distances travelled by Susan and David. Solve it to find the value of t.

14. Think of a number, double it, add 9 to the result, multiply this result by three, subtract 21 and divide by 6. The result is one more than the number that you first thought of.

(a) Let the number you thought of be x. What would it be after you had doubled it?

(b) What would it now be after adding 9?

(c) Now multiply the result by three.

(d) Taking the expression you arrive at in (c), subtract 21 from this.

(e) Now divide the whole expression by 6.

(f) By simplifying your answer to (e), check that your result comes to one more than x.

Inequalities

15. Mr Dean starts with an income of £6000, which is increased by £800 each year. Mrs Wilkins starts with an income of £9000, which is increased by £300 each year.

(a) After t years, what will Mr Dean's income be, in terms of t?

(b) After t years, what will Mrs Wilkins's income be, in terms of t?

(c) Write down an inequality to show when Mr Dean's income becomes greater than Mrs Wilkins's.

(d) After how many years will this be true?

16. A publisher offers to pay an author either 10 pence for every copy sold after the first 500, or 5p for every copy sold. Let c be the number of copies sold. Assume $c > 500$.

(a) If the author accepts the publisher's *first* offer, for how many copies will she receive 10p, in terms of c?

(b) How much money will she receive if she accepts the first offer?

(c) If the author accepts the publisher's *second* offer, how much money will she receive for c copies sold?

(d) Suppose the author will receive less money if she accepts the second offer. Form an inequality.

(e) Use your inequality to find the least number of copies for which this is true.

17. Arthur is 40 years old, Brian's age is 15 and Clare is 12 years old.

(a) What will be their ages in y years' time?

(b) After y years the combined ages of Brian and Clare are greater than Arthur's age. Write down an inequality to show this.

(c) Use this inequality to find the least value of y for which this will be true.

(d) How old will each be then?

18. At a shooting range, John has to pay 4p for a miss and receives 12p for each hit. He has 27 shots altogether.

(a) If John has x hits, how many misses will he have, in terms of x?

(b) How much will he have to pay for his misses, in terms of x?

(c) How much will he win from all his hits, in terms of x?

(d) As he only has 30p to spend, write down an inequality and use it to find the least value of x he can afford. (*Hint:* remember that he is *spending* the 30p.)

(e) Find how much he wins and how much he loses for this least number of hits, and use these to check the inequality.

19. Two whole numbers add up to 126. Five sevenths of one number is less than three fifths of the other.

(a) Letting one number be x, what is the other, in terms of x?

(b) Express five sevenths of the first number in terms of x.

(c) Express three fifths of the other number in terms of x.

(d) Form an inequality for the two fractions of the numbers.

(e) Solve this to find the possible values of x, and hence the largest possible value.

20. Five boys take part in a relay run of at least 20 km. Each boy runs a distance according to his ability. Paul is asked to run x kilometres, Stephen $\frac{2}{5}$ km, John $3x$ km, Christopher $6x$ km and Andrew $\frac{1}{3}$ km.

(a) What is the total distance that the boys will run, in terms of x?

(b) Form an inequality involving the least total distance that they will have to run.

(c) Paul says that he will only run the least possible distance needed, and only if it is less than 2 km. Solve the inequality to find whether Paul will have to run more than 2 km.

Exercise 2.6 *Angle calculations*

1.

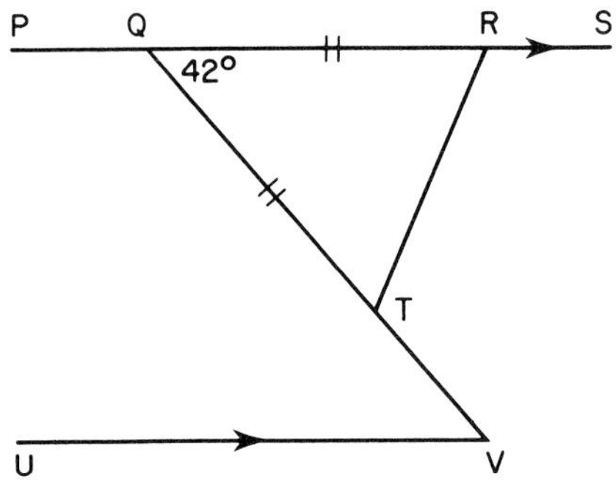

In the figure: PQRS is parallel to UV,
QR = QT,
$R\hat{Q}V = 42°$.
Calculate, giving reasons:

(**a**) $Q\hat{T}R$

(**b**) $S\hat{R}T$

(**c**) $U\hat{V}T$.

2.

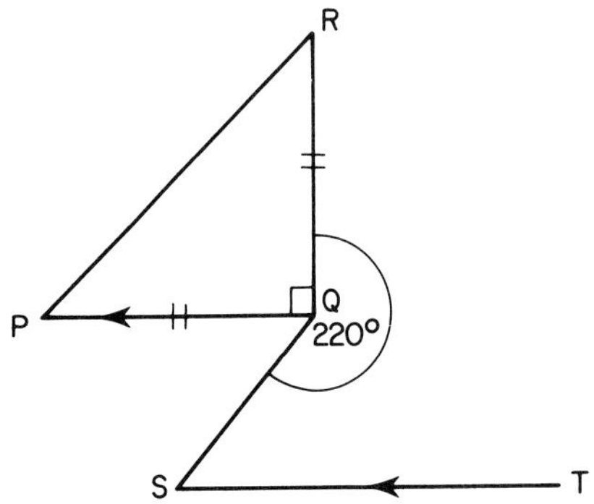

PQR is an isosceles triangle with a right angle at Q.
PQ is parallel to ST.
Reflex $R\hat{Q}S = 220°$.
Calculate, giving reasons:

(**a**) $P\hat{R}Q$

(**b**) $P\hat{Q}S$

(**c**) Reflex $T\hat{S}Q$.

3.

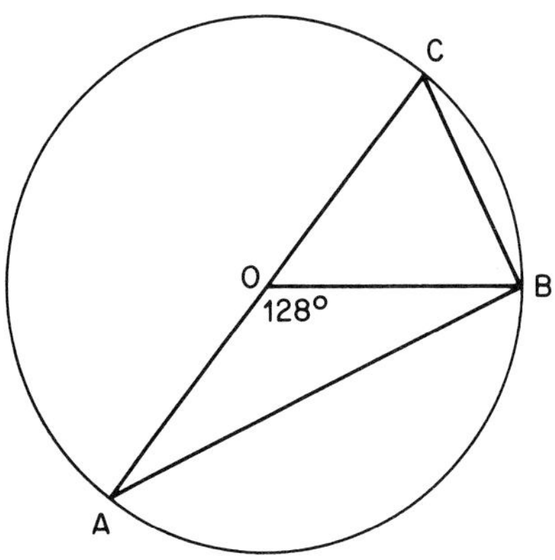

O is the centre of the circle with AOC as diameter.
$A\hat{O}B = 128°$.
Calculate, giving reasons:

(**a**) $O\hat{B}A$

(**b**) $O\hat{C}B$.

4.

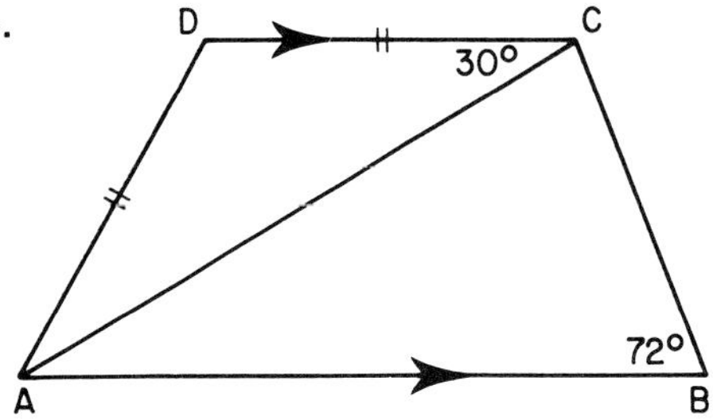

ABCD is a trapezium with DA = DC,
$A\hat{B}C = 72°$,
$A\hat{C}D = 30°$.
Calculate, giving reasons:

(**a**) $A\hat{C}B$

(**b**) $D\hat{A}B$.

5.

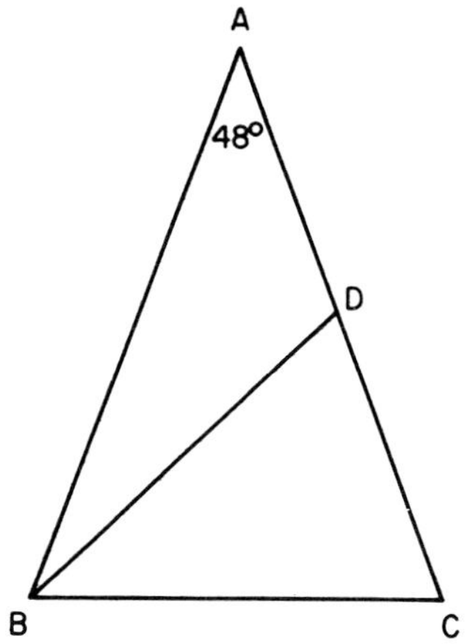

In the figure:
AB = AC,
BC = BD,
$B\hat{A}C = 48°$.
Calculate, giving reasons:

(**a**) $A\hat{C}B$
(**b**) $C\hat{B}D$
(**c**) $A\hat{D}B$.
(**d**) What is the relationship between $\triangle ABC$ and $\triangle BDC$?

6.

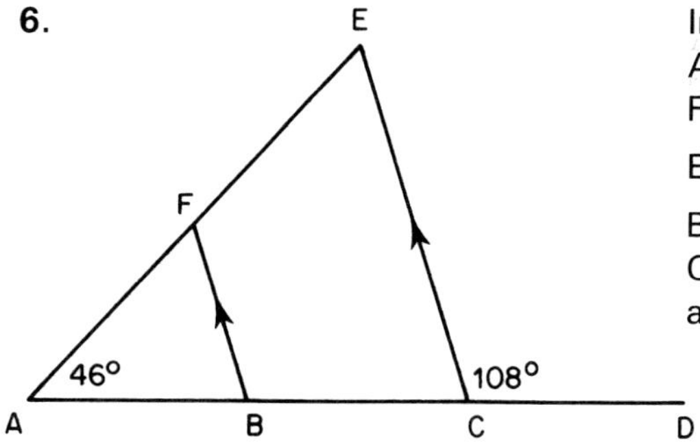

In the figure:
ABCD is a straight line,
FB is parallel to EC,
$E\hat{C}D = 108°$,
$B\hat{A}F = 46°$.
Calculate, giving reasons, the interior angles of the quadrilateral BCEF.

7.

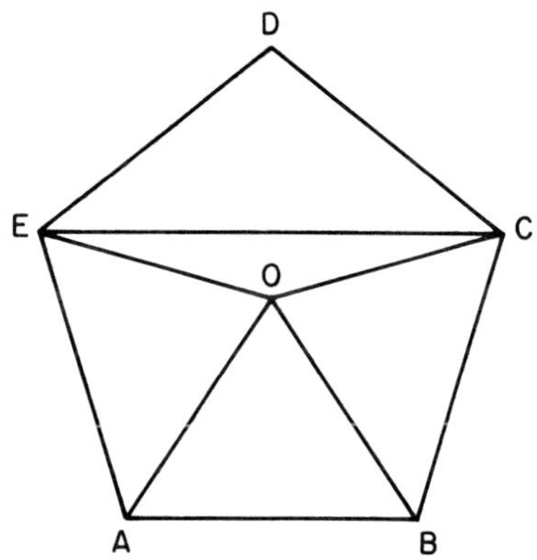

O is the centre of the regular pentagon ABCDE.
Calculate, giving reasons:

(**a**) $A\hat{O}B$
(**b**) $A\hat{B}O$
(**c**) $E\hat{D}C$
(**d**) $E\hat{C}D$
(**e**) $E\hat{O}C$
(**f**) $E\hat{C}O$.

8.

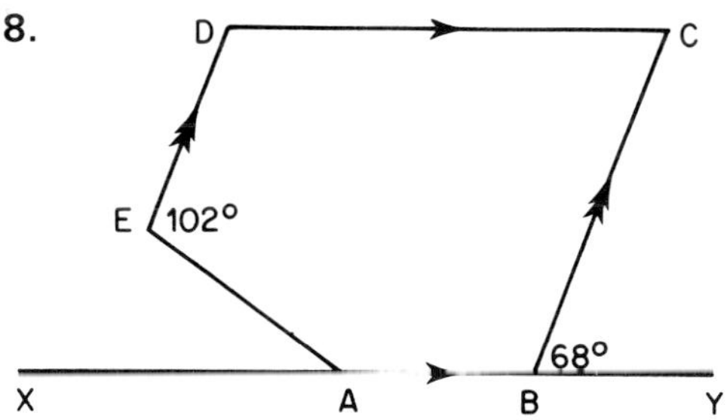

In the figure:
XABY is parallel to DC,
BC is parallel to ED,
$Y\hat{B}C = 68°$,
$D\hat{E}A = 102°$.
Calculate, giving reasons, the four unknown interior angles of the pentagon.

9.

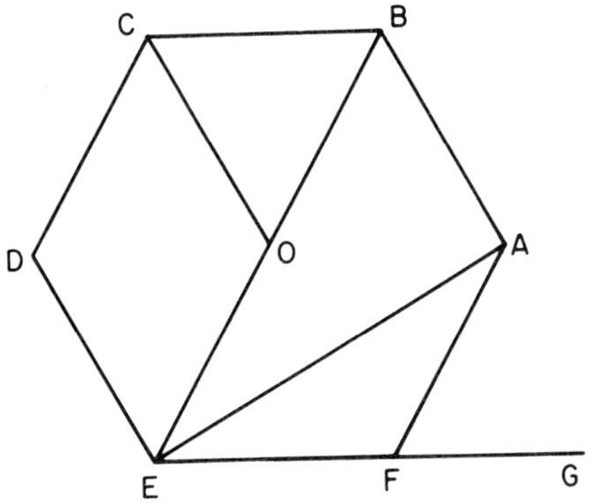

O is the centre of the regular hexagon ABCDEF.
EFG is a straight line.
Calculate, giving reasons:

(**a**) $\widehat{AFG}$
(**b**) $\widehat{AEF}$
(**c**) $\widehat{AEB}$
(**d**) $\widehat{BED}$
(**e**) $\widehat{OCB}$.

10.

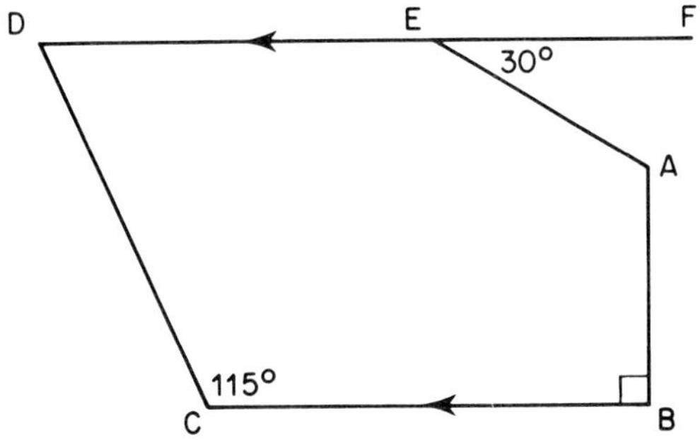

In the figure:
CB is parallel to DEF,
AB is perpendicular to CB,
$\widehat{FEA} = 30°$,
$\widehat{BCD} = 115°$.
Calculate, giving reasons:

(**a**) $\widehat{AED}$
(**b**) $\widehat{CDE}$
(**c**) $\widehat{BAE}$.

11.

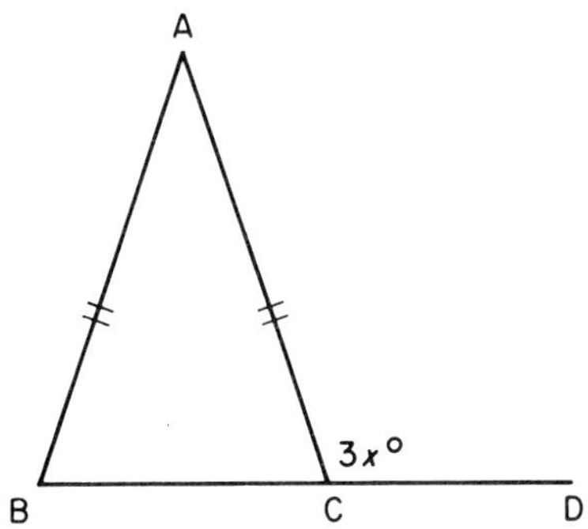

In the figure:
AB = AC,
BCD is a straight line,
$\widehat{CAB} = \frac{1}{2}\widehat{ABC}$,
$\widehat{ACD} = 3x°$.

(**a**) Write down the size of
(**i**) $\widehat{CAB}$ (**ii**) $\widehat{ABC}$ in terms of x.
(**b**) Form an equation and solve it for x.
(**c**) What is the size of $\widehat{ACD}$?

12.

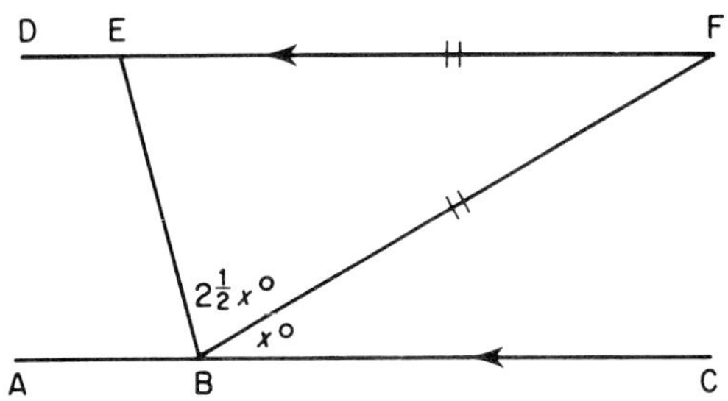

In the figure:
ABC is parallel to DEF,
FE = FB,
$\widehat{FBC} = x°$,
$\widehat{EBC} = 2\frac{1}{2}x°$.

(**a**) Write down the size of
(**i**) $\widehat{BEF}$ (**ii**) $\widehat{BFE}$ in terms of x.
(**b**) Form an equation and solve it for x.
(**c**) What is the size of $\widehat{ABE}$?

13.

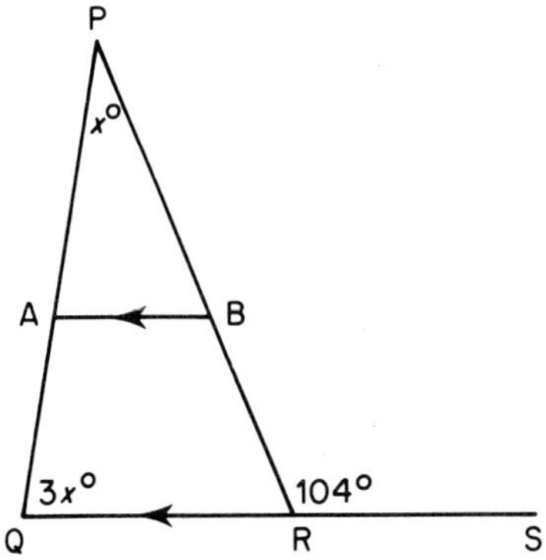

In the figure:
AB is parallel to QRS,
$Q\hat{P}R = x°$,
$P\hat{Q}R = 3x°$,
$P\hat{R}S = 104°$.
Find the value of x, and hence the size of $P\hat{A}B$.

14.

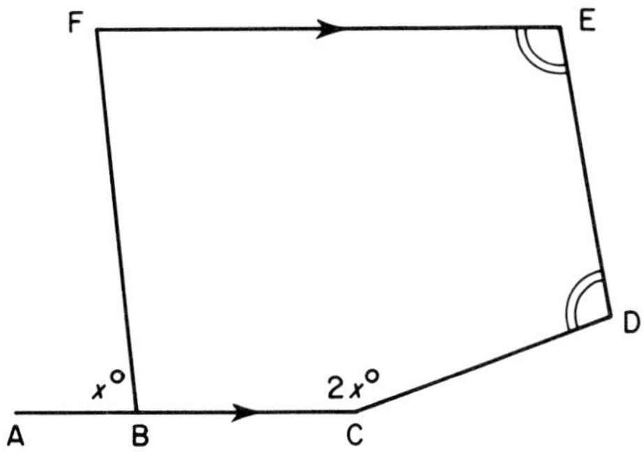

In the figure:
ABC is parallel to FE,
$F\hat{E}D = E\hat{D}C$,
reflex $F\hat{E}D = 260°$,
$A\hat{B}F = x°$,
$B\hat{C}D = 2x°$.

(**a**) What is the size of $F\hat{E}D$?
(**b**) What is the size of $E\hat{F}B$, in terms of x?
(**c**) What is the size of $C\hat{B}F$, in terms of x?
(**d**) Form an equation and find the value of x.

15.

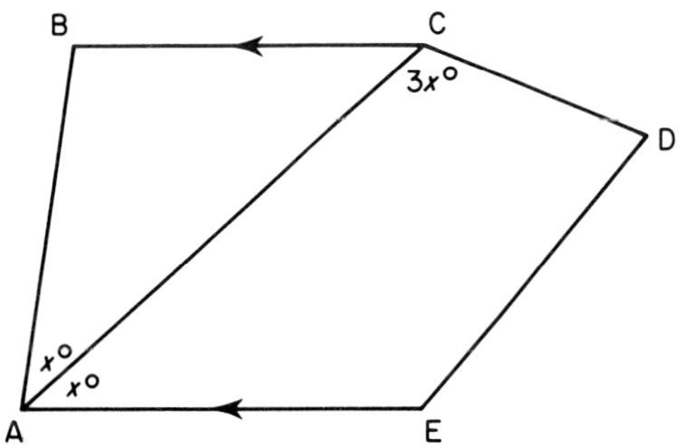

In the figure:
AE is parallel to BC,
AC bisects $B\hat{A}E$,
$A\hat{E}D + E\hat{D}C = 200°$,
$C\hat{A}E = x°$,
$A\hat{C}D = 3x°$.

(**a**) Calculate the value of x.
(**b**) Hence find the size of $A\hat{B}C$.
(**c**) What can you say about triangle ABC?

Reflection

1. Rectangle A has its vertices at (2, 1), (4, 1), (4, 4) and (2, 4).
- **(a)** Plot these points and join them to form rectangle A.
- **(b)** Rectangle B is the image of rectangle A after reflection in the line $y = -x$. Draw in rectangle B.
- **(c)** Rectangle C is the reflection of rectangle B in the line $y = 0$. Draw in rectangle C.
- **(d)** Rectangle D is the reflection of rectangle A in the line $y = 0$. Draw in rectangle D.
- **(e)** What mirror line could be used to reflect the rectangle C onto rectangle D?

2. Triangle ABC has its vertices at A(1, 2), B(0, 4) and C(3, 5).
- **(a)** Plot these points and join them to form triangle ABC.
- **(b)** Reflect triangle ABC in the line $y = 0$ to form the image A′B′C′.
- **(c)** Reflect triangle A′B′C′ in the line $y = x$ to form the image A″B″C″.
- **(d)** Give one point that is unchanged in position by the reflection in (c).

3. Mark scales from -5 to $+5$ on both x and y axes.
- **(a)** Plot the points A(-1, -3), B(-3, -1) and C(-4, -4) and join them to form triangle ABC.
- **(b)** Reflect triangle ABC in the line $x = 0$. Label this image triangle X.
- **(c)** Reflect triangle ABC in the line $y = 0$. Label this image triangle Y.
- **(d)** Triangle Z is the triangle required to complete the symmetry of the figure. Draw in triangle Z.
- **(e)** Reflection in what mirror line would map triangle ABC onto triangle Z?

4. Triangle A has its vertices at (-5, -2), (-3, 3) and (2, 5).
- **(a)** Plot these points and join them to form triangle A.
- **(b)** Reflect triangle A in the line $x = 0$ to form triangle B.
- **(c)** Reflect triangle B in the line $y = 0$ to form triangle C.
- **(d)** What mirror line could be used to reflect triangle A onto triangle C?
- **(e)** Draw in triangle D, the triangle needed to complete the symmetry of the figure.
- **(f)** What mirror line would be used to map triangle D onto triangle B?

5.
- **(a)** Copy the diagram, using a scale of 1 cm to 1 unit on both x and y axes.
- **(b)** Reflect triangle A in the line $y = x$ to give triangle B.
- **(c)** Reflect triangle A in the line $y = 0$ to give triangle C.
- **(d)** Reflect triangle B in the line $x = 0$ to give triangle D.
- **(e)** Reflect triangle C in the line $y = -x$ to give triangle E.
- **(f)** Reflect triangle D in the line $y = 0$ to give triangle F.
- **(g)** Reflect triangle A in the line $x = 0$ to give triangle G.

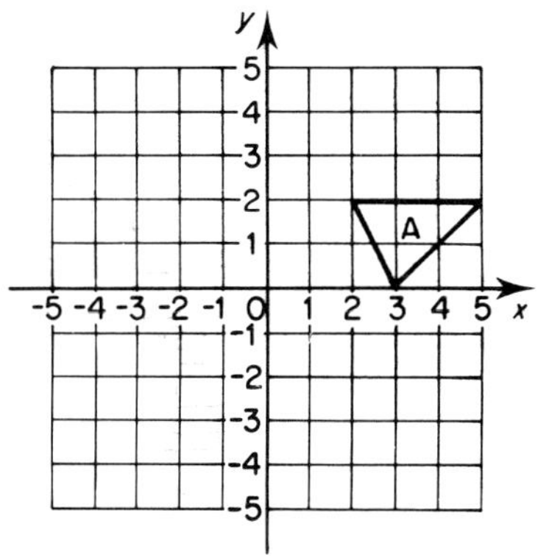

(h) H is the triangle required to complete the symmetry of the figure. In what mirror line would triangle C reflect onto triangle H?

(i) Which triangle would be reflected onto triangle H using the line $y = x$ as a mirror line?

6. Triangle ABC has vertices at A(−3, 1), B(−2, 4) and C(3, 2).

(a) Plot these points and join them to form the triangle ABC.

(b) Triangle ABC is reflected in the line $y = -x$ to give the triangle A′B′C′. Draw in triangle A′B′C′.

(c) Draw in the triangle A″B″C″, the image of A′B′C′ in the mirror line $x = 0$.

(d) Draw in the triangle A*B*C*, the triangle needed to complete the symmetry of the figure about the line $y = -x$.

7. Triangle ABC has its vertices at A(5, −5), B(−4, 1) and C(−1, 4).

(a) Plot these points and join them to form triangle ABC.

(b) Draw in the triangle A′B′C′, the image of ABC after reflection in $y = x$.

(c) Draw in the triangle A″B″C″, the image of ABC in the mirror line $x = 0$.

(d) Draw in triangle A*B*C*, the triangle that completes the symmetry of the figure.

(e) How many lines of symmetry has the complete figure?

8. Rectangle A has vertices at (1, −1), (4, 2), (2, 4) and (−1, 1).

(a) Plot these points and join them to form rectangle A.

(b) Reflect rectangle A in the line $x = 0$ to form rectangle B.

(c) Reflect rectangle A in the line $y = -x$ to form rectangle C.

(d) Rectangle D completes the symmetry of the figure.

(e) In what mirror line would rectangle B reflect onto rectangle D?

9. (a) Copy the diagram, using a scale of 1 cm to 1 unit on both x and y axes.

(b) Reflect triangle A in the line $y = 0$ to give triangle B.

(c) Reflect triangle B in the line $y = -x$ to give triangle C.

(d) Reflect triangle A in the line $y = -x$ to give triangle D.

(e) In what mirror line would triangle C reflect onto triangle D?

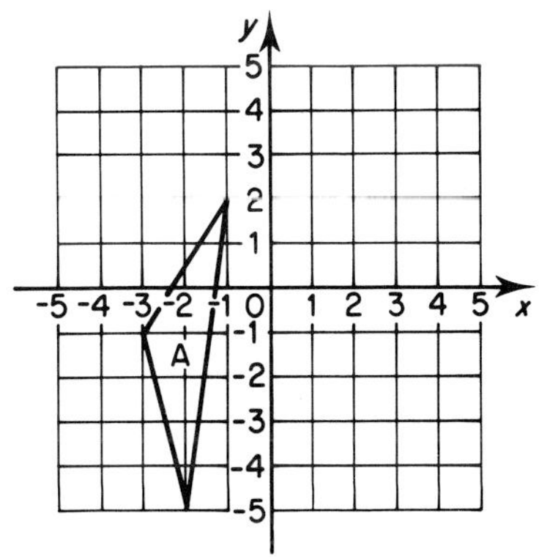

10. (**a**) Copy the diagram, using a scale of 1 cm to 1 unit on both x and y axes.

(**b**) Reflect quadrilateral A in the line $x = 0$ to give quadrilateral B.

(**c**) Reflect quadrilateral B in the line $y = 0$ to give quadrilateral C.

(**d**) Quadrilateral D is the quadrilateral required to complete the symmetry of the figure. In what mirror line would quadrilateral C reflect onto quadrilateral D?

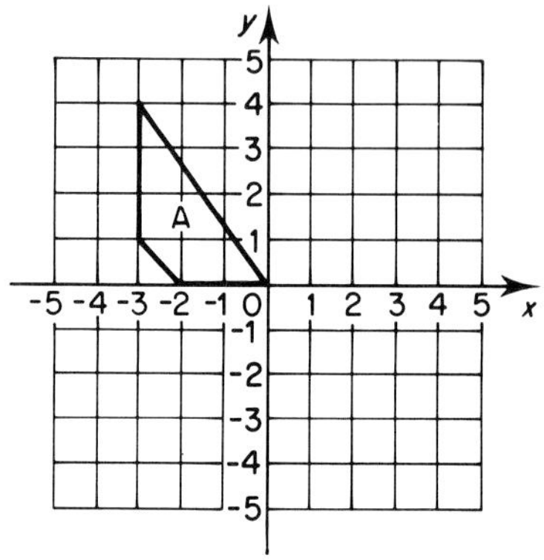

Rotation

11.

(**a**) What rotation would map triangle A onto triangle B?

(**b**) What rotation would map triangle D onto triangle C?

(**c**) What rotation would map triangle C onto triangle B?

(**d**) What anti-clockwise rotation would map triangle A onto triangle D?

(**e**) What order of rotational symmetry has the figure?

(**f**) How many lines of symmetry has the figure?

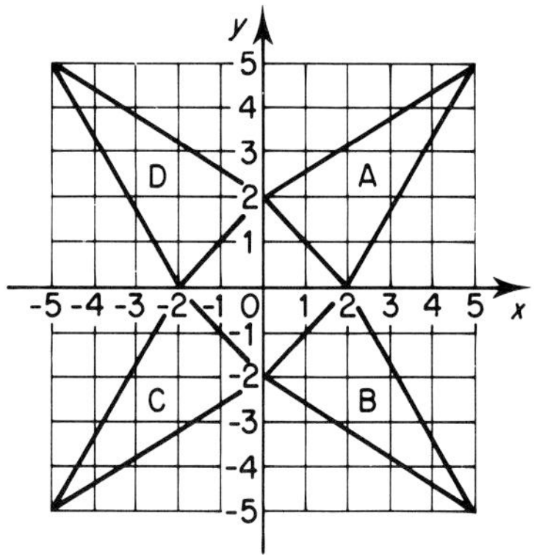

12. Triangle A has its vertices at $(-3, -3)$, $(-1, 3)$, $(-3, 3)$.

(**a**) Plot these points and join them to form triangle A.

(**b**) Rotate triangle A 180° about (0, 0) to form triangle B.

(**c**) Draw in triangle C, the triangle that, when rotated 90° anti-clockwise about (0, 0), maps onto triangle A.

(**d**) Triangle D completes the rotational symmetry of the figure. Draw in triangle D.

(**e**) Give one point that is unchanged in position by any of these rotations.

13. Triangle ABC has its vertices at A(3, 0), B(2, 4) and C(-4, 2).

(**a**) Plot these points and join them to form triangle ABC.

(**b**) Triangle A′B′C′ is the image of triangle ABC after a rotation of 90° about the point (2, 0). Draw in triangle A′B′C′.

(**c**) Triangle A″B″C″ is the image of triangle ABC after a rotation of 90° about the point (3, 2). Draw in triangle A″B″C″.

(**d**) What single transformation would map triangle A′B′C′ onto A″B″C″?

14. (a) Copy the diagram, using a scale of 1 cm to 1 unit on both x and y axes.
(b) What kind of quadrilateral is ABCD?
(c) Rotate ABCD 180° about the point (0, 0) to give the image A′B′C′D′.
(d) Which point will map onto A after this rotation?
(e) What kind of a quadrilateral is D′CDC′?
(f) What is the area of D′CDC′?
(g) What is the area of ABCD?

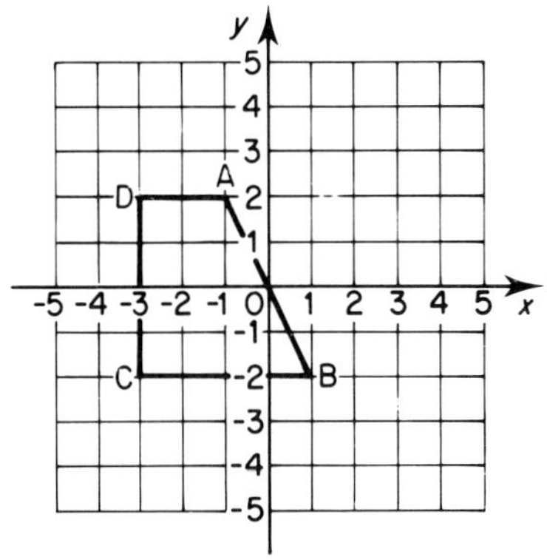

15. Triangle A has its vertices at (1, 3), (4, 2) and (4, 4).
(a) Plot these points and join them to form triangle A.
(b) Triangle B is the triangle that when rotated 180° about (0, 0) maps onto triangle A. Draw in triangle B.
(c) Triangle C is rotated 90° clockwise about (0, 0) to map it onto triangle A. Draw in triangle C.
(d) Triangle D completes the rotational symmetry of the figure. Draw in triangle D.
(e) Does the complete figure have line symmetry?

16. Triangle A has its vertices at (0, −2), (3, 5) and (−3, 5).
(a) Plot these points and join them to form triangle A.
(b) Triangle A is rotated 90° clockwise about (0, 0) to form triangle B. Draw in triangle B.
(c) Triangle C is the triangle that when rotated through 180° about (0, 0) maps onto triangle B. Draw in triangle C.
(d) Triangle D is the triangle that when rotated through 90° anti-clockwise about (0, 0) maps onto triangle B. Draw in triangle D.
(e) Shade in the area common to all the triangles.
(f) What kind of figure is the shaded area?

17. Triangle ABC has its vertices at A(4, 1), B(4, −1) and C(−4, 1).
(a) Plot these points and join them to form triangle ABC.
(b) Rotate triangle ABC 180° about (0, 0) to give triangle A′B′C′.
(c) What shape is the quadrilateral ABA′C?
(d) Rotate triangle ABC 180° about the point (0, 1) to give triangle A″B″C″.
(e) What shape is BCB″A?
(f) What is the area of BCB″A?

18. Quadrilateral A has its vertices at (1, 1), (0, 4), (−2, 4) and (−2, 3).
(a) Plot these points and join them to form quadrilateral A.
(b) Quadrilateral A is rotated 90° anti-clockwise about (0, 2) to give quadrilateral B. Draw in B.
(c) Quadrilateral B is rotated 180° about (1, −1) to give quadrilateral C. Draw in C.
(d) What single transformation would map quadrilateral C onto quadrilateral A?

19. (**a**) What rotation about which point would map triangle A onto triangle B?
(**b**) What rotation about which point would map triangle B onto triangle D?
(**c**) What rotation about which point would map triangle A onto triangle C?
(**d**) What single transformation would map triangle B onto triangle C?

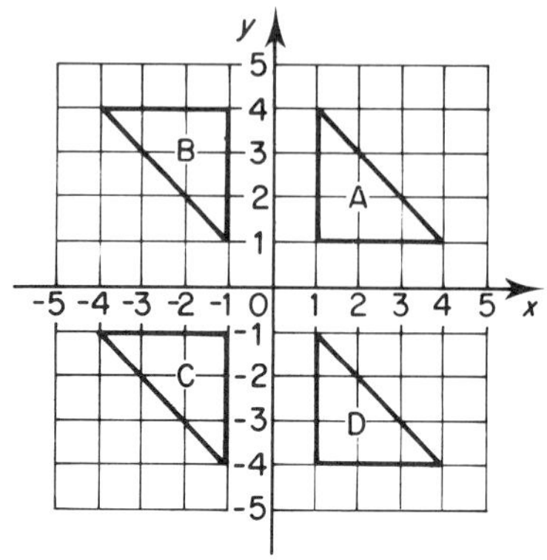

20. (**a**) Find three centres of rotation that could be used to map square A onto square B.
(**b**) Give the angle of rotation in each case.

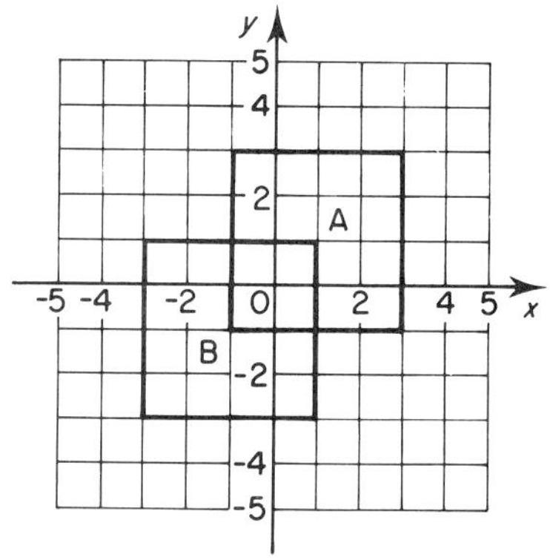

Translation

21. (**a**) What is the vector of the translation mapping triangle A onto triangle B?
(**b**) What vector would be required to translate triangle C onto triangle B?
(**c**) What vector would be required to translate triangle C onto triangle A?
(**d**) How is this related to the vectors in (a) and (b)?

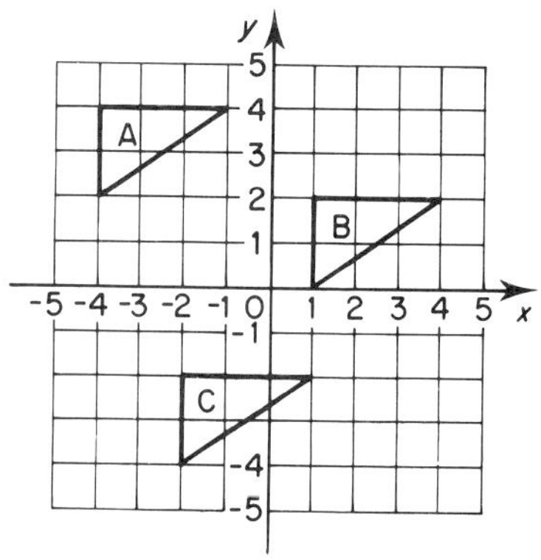

22. (a) What column vector describes the translation of triangle A onto triangle B?

(b) What column vector describes the translation of triangle C onto triangle D?

(c) What column vector describes the translation of triangle E onto triangle C?

(d) What column vector describes the translation of triangle A onto triangle D?

(e) What column vector describes the translation of triangle B onto triangle E?

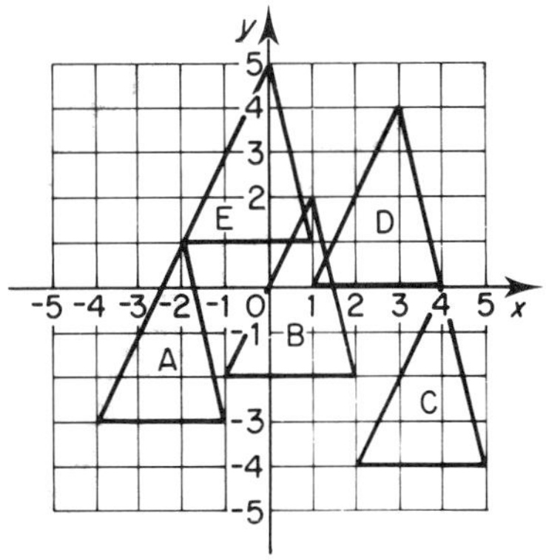

23. Triangle PQR has its vertices at P(3, 5), Q(4, 2) and R(1, 1).

(a) Plot these points and join them to form triangle PQR.

(b) Draw in the triangle P′Q′R′ formed by translating PQR by the vector $\begin{pmatrix} -5 \\ -2 \end{pmatrix}$.

(c) Translate triangle P′Q′R′ by the vector $\begin{pmatrix} 4 \\ -3 \end{pmatrix}$ to give the triangle P″Q″R″.

(d) What column vector describes the translation mapping triangle PQR onto P″Q″R″?

(e) How is this vector related to the vectors in parts (b) and (c)?

24. (a) What translation would map rectangle A onto D?

(b) What translation would map rectangle C onto G?

(c) What translation would map rectangle F onto B?

(d) What translation would map rectangle B onto E?

(e) What translation would map rectangle G onto A?

(f) What translation would map rectangle E onto F?

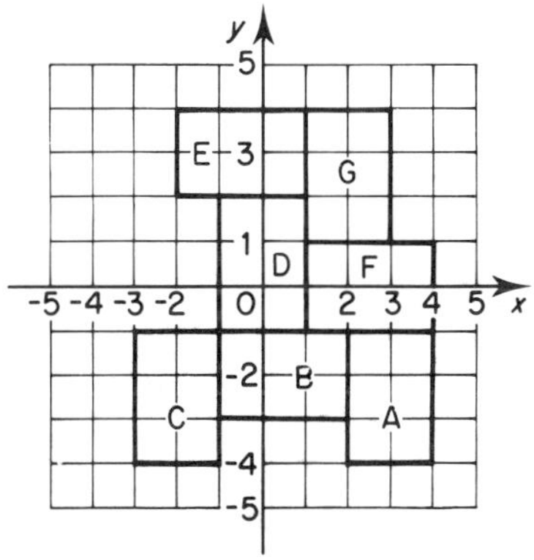

25. (a) By what vector must square A be translated to map onto square C?
(b) By what vector must square D be translated to map onto square I?
(c) Onto which square would square B map, after a translation of $\begin{pmatrix} 4 \\ 4 \end{pmatrix}$?
(d) Onto which square would square C map, after a translation of $\begin{pmatrix} -4 \\ 4 \end{pmatrix}$?

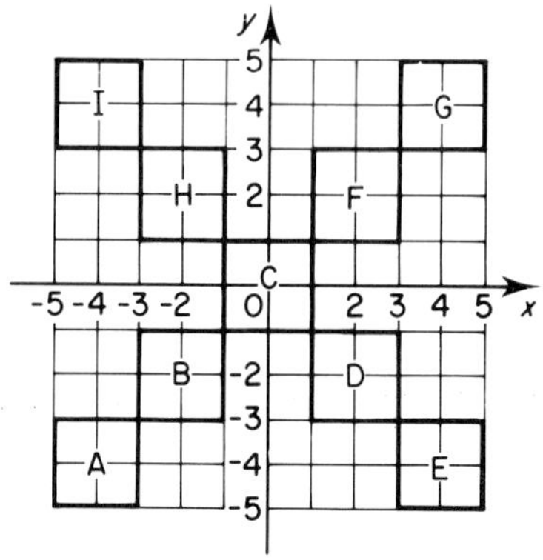

Combined transformations

26. (a) Copy the diagram, using a scale of 1 cm to 1 unit on both x and y axes.
(b) Rotate square A 90° about the point (1, 0). Label the image B.
(c) Translate square A by the vector $\begin{pmatrix} 2 \\ -4 \end{pmatrix}$. Label this image C.
(d) What translation would map square C onto square B?
(e) Give the details of another transformation that would map square C onto square B.

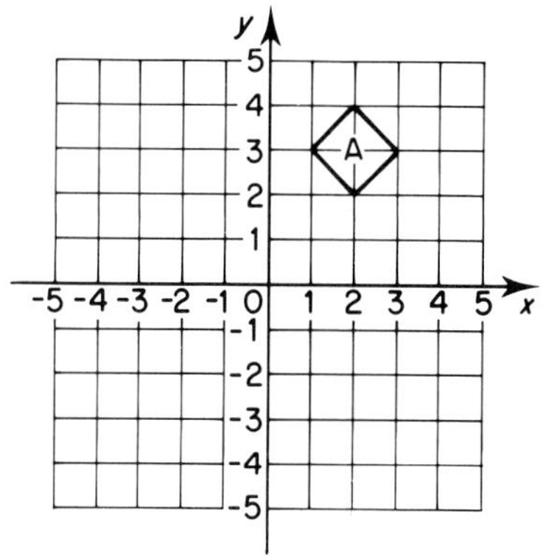

27. Triangle A vertices (1, 1), (3, 2), (2, −3).
(a) Plot these points and join them to form triangle A.
(b) Rotate triangle A 90° anti-clockwise about the point (0, 2). Letter this image B.
(c) Translate triangle A by the vector $\begin{pmatrix} -5 \\ -3 \end{pmatrix}$. Letter this image C.
(d) Translate triangle B by the vector $\begin{pmatrix} -5 \\ -3 \end{pmatrix}$. Letter this image D.
(e) What centre of rotation would map triangle C onto triangle D, and what angle would be required?

28. Triangle A has its vertices at $(-1, 2)$, $(-1, -4)$ and $(2, -4)$.

(a) Plot these points and join them to form triangle A.

(b) Triangle A is reflected in the line $y = x$ to give triangle B.

(c) Triangle B is reflected in the line $y = -1$ to give triangle C.

(d) Through what angle and about which point could triangle C be rotated onto triangle A?

29. (a) Copy the diagram, using a scale of 1 cm to 1 unit on both x and y axes.

(b) Square A is translated by the vector $\begin{pmatrix} -2 \\ -1 \end{pmatrix}$. Draw its image under this translation and letter it B.

(c) Give the centre of the 180° rotation that would map square B onto square A.

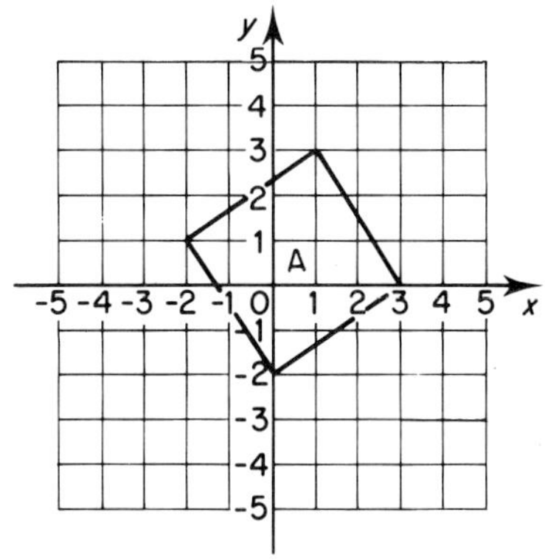

30. (a) Copy the diagram, using a scale of 1 cm to 1 unit on both x and y axes.

(b) Triangle B is the image of triangle A after reflection in the line $x = 0$. Draw in triangle B.

(c) Triangle C is the image of triangle A after reflection in the line $y = 0$. Draw in triangle C.

(d) What single transformation would map triangle B onto triangle C?

(e) Draw in the fourth triangle, D, that would complete the symmetry of the figure.

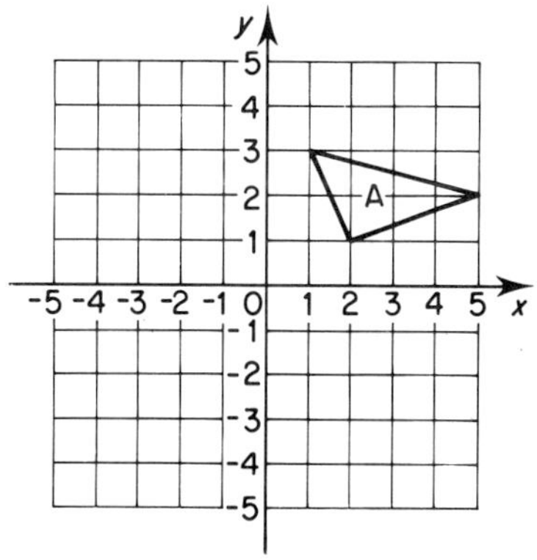

31. Triangle A has vertices at (1, 1), (1, 4), (3, 3).
  (a) Plot these points and join them to form triangle A.
  (b) Triangle A is rotated 90° clockwise about the point (1, 1) to give triangle B.
  (c) Triangle B is rotated 90° anti-clockwise about the point (0, −2) to give triangle C.
  (d) What translation would map triangle C onto triangle A?

32. The quadrilateral ABCD has vertices at A(−4, 3), B(1, 3), C(3, 1) and D(−2, 1).
  (a) Plot these points and join them to form the quadrilateral ABCD.
  (b) What kind of quadrilateral is ABCD?
  (c) Reflect ABCD in the line $y = x$ and label the image A′B′C′D′.
  (d) Join DD′. What kind of quadrilateral is ABCA′?
  (e) What transformation would map triangle ABD′ onto DCA′? (Be careful!)

33. Triangle ABC has vertices at A(4, 5), B(4, 2), C(−4, 5).
  (a) Plot these points and join them to form triangle ABC.
  (b) Reflect triangle ABC in the line $y = x$ to give triangle A′B′C′.
  (c) Give two points that are unchanged by this reflection.
  (d) Reflect triangle ABC in the line $y = 0$ to give triangle A″B″C″.
  (e) Are there any points of triangle ABC unchanged by this reflection?
  (f) Triangle A″B″C″ is reflected in the line $y = x$ to give triangle A*B*C*.
  (g) What transformation would map triangle A*B*C* onto A′B′C′?

Exercise 2.8 *Ratio and percentage*

1. There are 162 cars and 27 motorcycles in a car-park. Express the number of cars to the number of motorcycles as a ratio in lowest terms.

2. A square has side 8 cm and a rectangle measures 12 cm by 6 cm. Write
  (a) the ratio of their perimeters in lowest terms
  (b) the ratio of their areas in lowest terms.

3. The ages of Jane and John are in the ratio 3 : 2. How old is Jane when John is 6 years old?

4. AB and CD are two lines whose lengths are in the ratio 7 : 4. Find the length of AB if CD = 10 cm.

5. Bill and Barbara share a bowl of cherries in the ratio of 7 : 3. How many cherries are there in the bowl if Barbara's share is 21?

6. The angles of a triangle are in the ratio 16 : 11 : 9. Calculate the size of the angles.

7. George, Alison and Peter share £35 in the ratio 5 : 3 : 2. How much does Alison receive?

8. Increase 99 in the ratio of 11 : 9.

9. Charlie's pocket money was 60p before it was increased in the ratio of 5 : 4. What is it now?

10. Decrease 63 in the ratio of 3 : 7.

11. 7 bottles of 'Fizzy' cost £2.24. What is the cost of 3 bottles?

12. The number of kilograms of potatoes, carrots and onions sold by a greengrocer in a day is in the ratio 5 : 2 : 1.

(**a**) 14 kg of carrots are sold. Find out how many kilograms of potatoes and onions are sold.

The price per kilogram is in the ratio 1 : 3 : 2.

(**b**) Onions cost 24p per kilogram. Calculate the total takings from potatoes, carrots and onions.

13. A box is 30 cm long, 24 cm high and 8 cm wide.

(**a**) Calculate its volume.

Another box is such that its length is less than that of the first box in the ratio of 9 : 10, the height is less in the ratio of 3 : 4 and the width is greater in the ratio of 5 : 4.

(**b**) Calculate the dimensions of the new box and hence find its volume.

(**c**) Compare the volumes of the two boxes as a ratio in lowest terms.

14. The number of dolls owned by Angela, Brenda and Carol is in the ratio of 5 : 3 : 1.

(**a**) If Carol has x dolls, how many dolls do Angela and Brenda have, in terms of x?

(**b**) Each girl is given another 2 dolls for Christmas. Write down an expression for the total number of dolls the girls have now.

(**c**) Altogether they now own 24 dolls. Form an equation and solve it for x.

(**d**) How many dolls did Brenda start with?

15. Find (**a**) 82% of 400 (**b**) 8% of £3.50 (**c**) 18% of 3.5 m, in centimetres.

16. What is (**a**) 27 as a percentage of 90

(**b**) 42p as a percentage of £2.10

(**c**) 550 g as a percentage of 2.5 kg?

17. 160 shoppers were asked which soap they would buy.

45% chose Bodyclean
35% chose Superwash
15% chose Scrub
5% chose other brands.

How many chose each brand?

18. In an orchard of 600 trees there are 330 apple trees, 150 pear trees and the rest are plum trees. Write these numbers as a percentage of the total number of trees in the orchard.

19. A bucket and spade cost £2.50.

(**a**) The bucket costs 20% of the total. How much does it cost?

(**b**) Write the cost of the bucket as a percentage of the cost of the spade.

20. Richard pays £45 for his season ticket on the railway. What will it cost him when the fare is increased by 8%?

21. Mr Old, an antique dealer, buys a desk for £80. He sells it to Mr Robinson, making a profit of 35%.

(**a**) For how much does he sell it?

(**b**) Mr Robinson then sells it to a friend for £135. What is Mr Robinson's profit as a percentage of the price he paid for it?

22. In a school of 200, 80% are boys.

(**a**) How many boys and how many girls are there?

(**b**) The number of boys decreases by 5% and the number of girls increases by 10%. How many boys and girls are there now?

(**c**) Has the total number of pupils increased or decreased, and by what percentage?

23. The Mason family paid £40 000 for their house 4 years ago.

- **(a)** What was the house worth after 2 years, if the value had increased by 20%?
- **(b)** After another 2 years the value had dropped by 20%, because of the construction of a motorway. What is the house worth now?
- **(c)** Is the house worth more or less than the original purchase price, and by what percentage?

24. Mr Rolls buys a car for £6500.

- **(a)** After 27 months he sells it to a garage at a loss of 40%. How much did he sell it for?
- **(b)** The garage sells the car to Mr Royce, making a profit of 25% on what it had paid for it. How much did Mr Royce pay?
- **(c)** What percentage of the original price is the car worth now?

25.
- **(a)** Find $33\frac{1}{3}$% of £1.50.
- **(b)** A bookseller buys a book for £1.50 and sells it at a $33\frac{1}{3}$% profit. For how much does he sell the book?
- **(c)** He buys another book for £2.40 and sells it for £3.00. What is his percentage profit this time?

26. A potter makes some small saucers. Each one costs him 24p to make. He sells them to a shop, making a profit of 25%.

- **(a)** What is the profit on a saucer?
- **(b)** The shop makes a further 40% profit. What is the price of a saucer when the shop sells it?
- **(c)** Express the total profit as a percentage of the potter's manufacturing costs.

27. Mr Bailey pays 30% of his income in tax.

- **(a)** What does he pay in tax, if his income is £11 500?
- **(b)** If he paid £2700 in tax, what would his income be?

28. Value Added Tax (VAT) is calculated at a rate of 15%.

- **(a)** What is the amount of VAT on a knife worth £3 before tax?
- **(b)** What is the value of a bed, before tax, if the amount of VAT is £22.50?

29. A tourist pays 20p for a 32-page guide book.

- **(a)** What is the cost per page?
- **(b)** What is the cost per page when the price is increased by 5% and the number of pages is reduced in the ratio of 7 : 8?

30. Which is larger, 88 increased in the ratio of 12 : 11 or 80 increased by 20%?

3

Common Entrance Paper II: 30-minute tests

Within each exercise, a possible mark scheme is indicated in parentheses.

Exercise 3.1

1. The inside of a water tank is 30 cm high and has a rectangular base measuring 40 cm by 15 cm. It is filled with water to a depth of 20 cm.

(a) Find the volume of water in the tank
(**i**) in cubic centimetres (*5*) (**ii**) in litres. (*2*)

(b) First, three more litres of water is added. Find the depth of water in the tank. (*4*)

(c) A tap is then opened at the base of the tank and it drains completely in two minutes. Find the average rate of flow of the water, in cubic centimetres per second. (*5*)

2. The ancient fortress of Basilica had guards on each of its four walls. There were x soldiers on the North wall.

(a) There were twice as many soldiers on the South wall as on the North wall. Write down an expression for the number of soldiers on the South wall. (*1*)

(b) There were 5 fewer soldiers on the East wall than on the North wall. Write down an expression for the number of soldiers on the East wall. (*1*)

(c) Write down an expression for the difference between the number of soldiers on the South wall and the number on the East wall. Simplify your answer. (*3*)

(d) There were 25 soldiers on the West wall. Write down, and simplify, an expression for the number of soldiers guarding Basilica. (*4*)

(e) At any time, there were twice as many soldiers off duty as there were guarding the fortress. Altogether there were 276 soldiers. Write down an equation in x and solve it. (*7*)

(f) Which wall was guarded by most soldiers? (*1*)

3. In this question, do all constructions using ruler and compasses only.

(a) Construct a triangle ABC with base BC = 10.5 cm, CA = 7 cm and AB = 9 cm. Extend AC to D, such that CD is 6 cm. (*6*)

(b) Construct the perpendicular from A to BC and the bisector of angle BCD. Mark the point P where these two lines meet. (*6*)

(c) Measure angle BAC, angle BCD and AP. (*3*)

(d) Join BD and shade the set of points inside triangle BCD which are nearer to CD than CB. (*2*)

Exercise 3.2

1. In Figure 1, ABCDEF is a regular hexagon.
   (a) What kind of triangle is BCD? *(2)*
   (b) Calculate, giving reasons
      (i) angle BCD *(3)*
      (ii) angle CBD *(3)*
      (iii) angle ABD. *(3)*
   (c) What kind of quadrilateral is ABDE? Give reasons. *(3)*
   (d) Which diagonal of the hexagon is also an axis of symmetry of ABDE? *(2)*

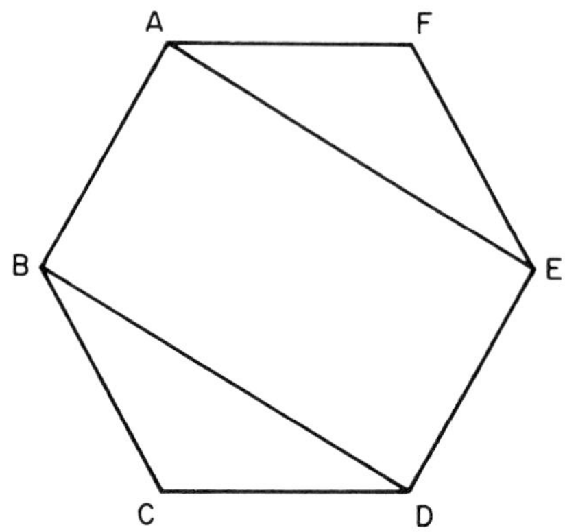

Figure 1

2. In a survey, 300 teenagers were asked the names of their favourite pop group.
   55 said Ultrapox
   45 said Crewcut 100
   50 said Mad Donna
   60 said Dead on Arrival
   90 did not like pop music.
   (a) Display this information as a pie chart. *(7)*
   (b) 75% of those who said Dead on Arrival thought their best single was 'You spin me round like a washing machine.' What percentage of all those surveyed was this? *(5)*
   (c) All the 90 who did not like pop music liked classical music. Of these,
      50% liked Mozart best
      40% liked Beethoven best
      10% liked some other composer best.
      (i) How many people liked Beethoven best? *(2)*
      (ii) Write, in its lowest terms, the ratio of this number to the number of all those in the survey. *(3)*
3. (a) Anderton is 180 km from Clarkville, with Bigsby on the road between the two and 120 km from Anderton. Using scales of 2 cm to 1 hour and 1 cm to 10 km, draw a diagram showing time from 0900 to 1700 on the horizontal axis and distance from Anderton on the vertical axis. Mark the positions of Bigsby and Clarkville on the vertical axis. *(3)*
   (b) Mrs Darnley sets out by car from Anderton at 0900, to visit a friend in Clarkville. She drives to Bigsby at 80 km/h, stops there for 15 minutes for coffee and continues on to Clarkville, arriving at 1200. Mark her journey on your diagram. *(5)*
   (c) At what speed did she travel for the second part of her journey? *(3)*
   (d) At a quarter to three, she suddenly notices the time and leaves promptly for home, driving at 90 km/h. Mark on your diagram her stay in Clarkville and her journey home. *(4)*
   (e) At what time does she pass through Bigsby and when does she arrive home? *(2)*

Exercise 3.3

1. Mr Haker wishes to hang wallpaper on a dining room wall 6 m long and 2.5 m high. Paper comes in rolls 75 cm wide and 15 m long, costing £5.50 each. Paper is to be hung in vertical strips, as in Figure 1.

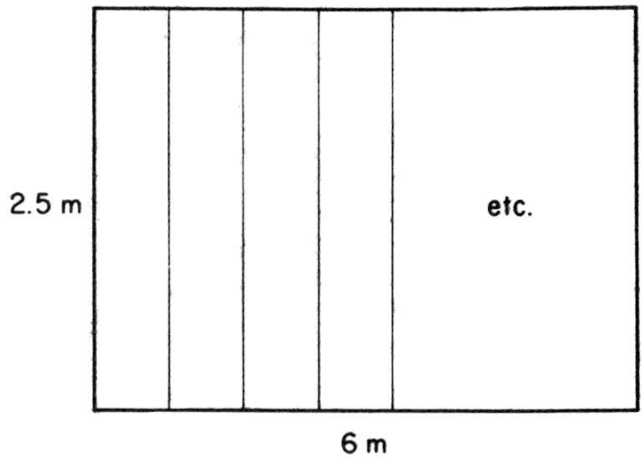

Figure 1

   (a) How many strips will he have to hang? *(4)*
   (b) How many rolls will he need and how much will they cost? *(4)*

   He takes 35 minutes to hang each strip, starting at 12.45 p.m. He wants to watch the international rugby match, which starts at 3 p.m.

   (c) Find how many strips he can hang before the match begins and at what time he will finish hanging these strips. *(5)*
   (d) After the match, he starts work again at 5 p.m. and continues at the same rate as before until he finishes. At what time will he finish? *(4)*

2. (a) Lady Armitage buys a ring for £350. She later sells it to the Countess of Osnabrück for £490. Find her percentage profit. *(4)*
   (b) When the Countess is in debt she has to sell the ring at a 30% loss. For how much does she sell it? *(4)*
   (c) Write the price for which the Countess sold the ring as a percentage of the price for which Lady Armitage bought it. *(4)*
   (d) Also write the ratio
   £350 : £490 : the Countess's selling price
   in its lowest terms. *(4)*

3. X and Y are two points on a stretch of coast running North–South, with sea to the East. X is 10.6 km North of Y. Z is a lightship 9.2 km from Y on a bearing of 030°.
   (a) Using a scale of 1 cm to 1 km, make a drawing showing X, Y and Z. *(4)*
   (b) P is a point on the coast between X and Y, 6.4 km from X. A boat sets out from P, sailing due East. Q is the point where its course meets YZ. Mark its journey from P to Q on your drawing. *(3)*
   (c) It takes 20 minutes to reach Q. Calculate its average speed between P and Q. *(3)*
   (d) When it reaches Q, a mist comes down, so that visibility everywhere is reduced to 3 km. Draw circles to show the limits of visibility from Q and Z. *(4)*
   (e) Shade the set of points inside the triangle XYZ which are visible from both Q and Z. *(3)*

Exercise 3.4

1. In Figure 1, ABCDEF is a regular hexagon with centre O.
   (a) What kinds of triangle are
   (i) OFE *(1)* (ii) OCE? *(1)*
   (b) Find, giving reasons,
   (i) angle EOC *(3)*
   (ii) angle OCE *(3)*
   (iii) angle FEC *(3)*
   (iv) angle FED (remember, it is one of the angles of ABCDEF) *(3)*
   (v) angle CED. *(1)*
   (c) Give a reason why FC is parallel to ED. *(2)*

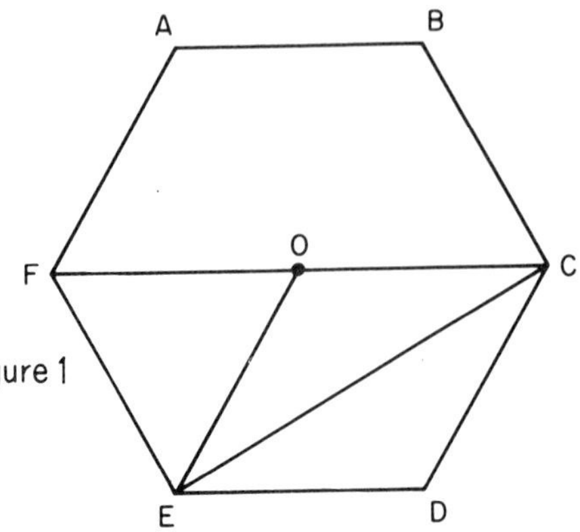

Figure 1

2. The members of a class of 30 girls were asked if they enjoyed knitting or netball.
   1 enjoyed neither,
   22 enjoyed netball.
   (a) Use a Venn diagram to find how many enjoyed knitting but not netball. *(5)*
   (b) 18 enjoyed knitting. How many enjoyed both? *(3)*
   (c) Further enquiry revealed that 20 of the girls enjoyed lacrosse. Draw two Venn diagrams to show the maximum and minimum numbers who could have enjoyed all three activities. *(9)*
3. When Mr Eagle went on holiday in Spain, £1 was equivalent to 210 pesetas.
   (a) Draw a conversion graph suitable for converting amounts up to £2 to pesetas. *(4)*
   (b) Use your graph to find, to the nearest whole number:
   (i) £1.20 in pesetas *(2)*
   (ii) 27 pesetas in pence *(2)*
   (iii) 90p in pesetas *(2)*
   (iv) 87 pesetas in pence. *(2)*
   (c) Mr Eagle changed £150 into pesetas. He was charged 400 pesetas commission. Use your graph to find, to the nearest 100 pesetas, how much he received. *(4)*

Exercise 3.5

1. (**a**) What is the total surface area of the solid in Figure 1? (*9*)
   (**b**) The solid is to be built out of cubes, each having an edge of length 0.5 cm. How many cubes are needed? (*8*)

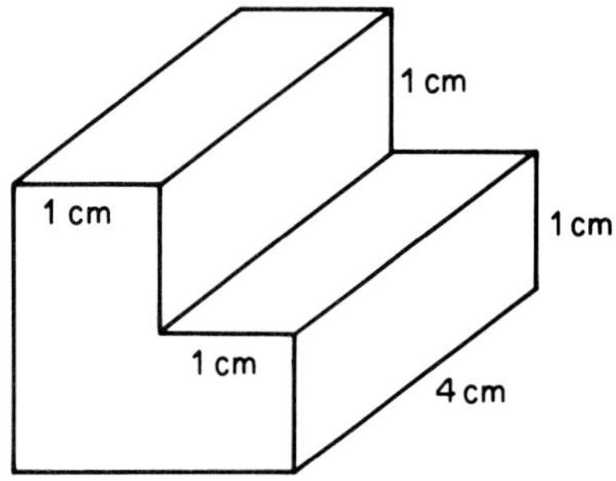

Figure 1

2. Muckleton Farm is 30 km from Bridlem. At 1330, Tom leaves the farm, travelling at 18 km/h on the tractor, for Diggle's Field, 12 km from Muckleton on the Bridlem Road. When he arrives, Tom takes 30 minutes to unload the trailer but then finds he is unable to restart the tractor.
   (**a**) Using a scale of 1 cm to 2 km on the vertical axis and 1 cm to 20 minutes on the horizontal axis, draw a travel graph to show Tom's movements until he has finished unloading the trailer. (*7*)
   (**b**) At what time does Tom get to Diggle's field? (*1*)
   His boss, Farmer Potter, leaves Bridlem at 1430 and travels towards the farm in his car, at 54 km/h. As he passes Diggle's field, he notices Tom sitting by the gate.
   (**c**) Mark on your graph Farmer Potter's journey to Diggle's Field, and Tom's wait. (*3*)
   (**d**) For how long has Tom been waiting when Farmer Potter arrives? (*1*)
   Farmer Potter gets out and tries to restart the tractor. After half an hour he gives up and decides to tow the tractor back to the farm, which he does at 8 km/h, taking Tom with him.
   (**e**) Complete the graph of their movements. (*3*)
   (**f**) At what time do they arrive at the farm? (*1*)
3. Kim and Sheila go to the shops and buy x bars of chocolate at 3 pence each and $(x - 3)$ packets of crisps at 15 pence each.
   (**a**) How much did they spend on chocolate, in terms of x? (*3*)
   (**b**) How much did they spend on crisps, in terms of x? (*3*)
   (**c**) How much did they spend altogether, in terms of x? (*2*)
   (**d**) Kim had 26p and Sheila had 34p. Write down an inequality relating what they spent to what they could afford to spend. (*5*)
   (**e**) Solve the inequality and so find the largest number of bars of chocolate and packets of crisps that they could buy. (*4*)

Exercise 3.6

1. **(a)** Construct a rhombus with diagonals of length 9 cm and 7 cm. Measure the length of one side of the rhombus. *(7)*
   (b) Construct a triangle ABC, in which angle A = 114°, AB = 8.3 cm and AC = 6.3 cm. Measure BC. *(5)*
   (c) In the same triangle, construct the perpendicular from B to CA produced. Mark the point X where this line meets CA produced. Measure BX. *(5)*
2. **(a)** Six farmers paid a bill for machinery. Three of them paid one eighth of the bill each; the others shared the rest of the bill equally. What fraction of the bill was paid by each of the other three farmers? *(5)*
   (b) The total bill was £6240. How much did each of the six farmers pay? *(6)*
   (c) If the farmers had not paid then, they would have been charged interest at 12% per year. How much would the total bill have been after one year? *(5)*
3. In Figure 1, XYZ is a triangle with angle Y = 48°.
   XP is parallel to YQ,
   XQ bisects the angle ZXP,
   XQ meets YZ produced at Q,
   Angle XQZ = 42°.
   (a) Calculate, giving reasons:
   (i) angle QXP *(2)*
   (ii) angle XZQ *(4)*
   (iii) angle YXZ. *(4)*
   (b) The figure XQZY is reflected in the line XQ. S is the image of Y, T is the image of Z.
   (i) On which line will T lie? *(3)*
   (ii) Give a reason why YXS is a straight line. *(4)*

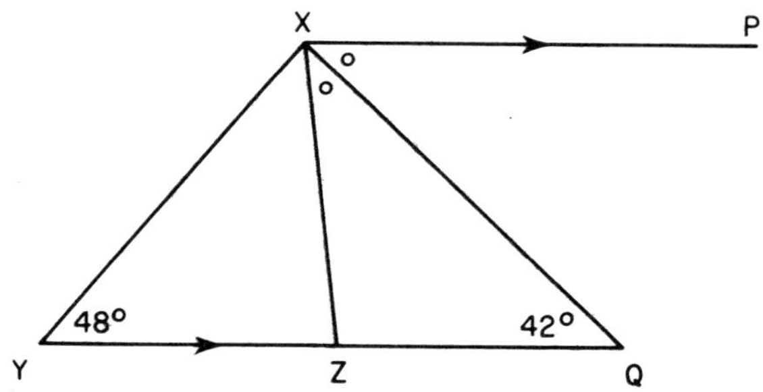

Figure 1

Exercise 3.7

1. A rectangle has a length of 24 cm and a width of 12 cm.

(a) Find the area of the rectangle. *(4)*

(b) Find the area of a square whose perimeter is the same as that of the rectangle. *(4)*

(c) Which is larger in area, and by how much? *(2)*

(d) Express the area of the rectangle as a fraction of that of the square, in its lowest terms. *(3)*

(e) Express the difference in areas as a percentage of the area of the rectangle. *(4)*

2. ABCD is a rectangular field, with B 160 m due East of A and D 110 m due North of A.

(a) Using a scale of 1 cm to 10 m, draw a plan of the field. *(3)*

(b) A tree is equidistant from AB and AD, and also equidistant from AD and BC. Show, by drawing, the position of the tree in the field. Letter this point T. *(5)*

(c) There is treasure buried 40 m from the tree on a bearing of 154° from the tree. Show, by drawing, the position of the treasure. *(6)*

(d) Find, by measurement, the distance of the treasure from the points A, B and C respectively. *(3)*

3. (a) A man walks at 5 km/h and runs at 15 km/h. He covers a distance of 22 km in 2 hours, partly walking and partly running. If the distance he walked was x km, find, in terms of x:

(i) the distance he ran *(2)*

(ii) how long he took to run this distance *(2)*

(iii) how long it took him to walk the x km. *(2)*

(iv) Form an equation for the total time. Solve it to find x. *(5)*

(b) If $\frac{1}{v} - \frac{1}{u} = \frac{1}{f}$, find f when $v = 3\frac{1}{2}$, $u = 4\frac{2}{3}$. *(5)*

Exercise 3.8

1. Figure 1 shows a regular hexagon ABCDEF, with centre O.

(a) Describe fully:

(i) the transformation which maps FOE onto ABO (*4*)

(ii) the transformation which maps FEDO onto FABO (*3*)

(iii) the transformation which maps FCBA onto ADEF. (*4*)

(b) **(i)** What kind of quadrilateral is FEDO? (*2*)

(ii) Write down the axes of symmetry of FEDO or show them on a rough sketch. (*3*)

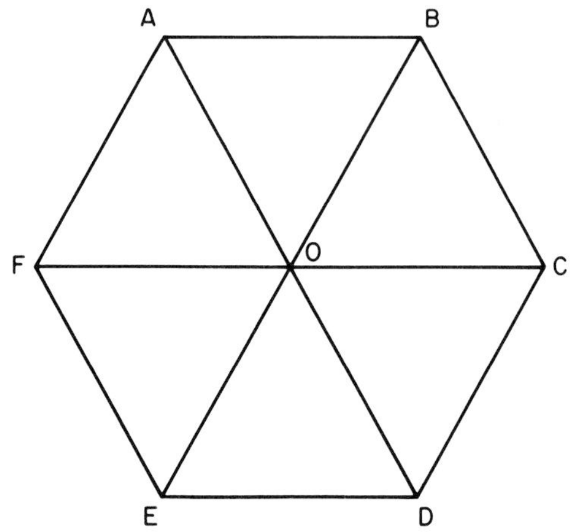

Figure 1

2. 600 people were asked which make of crisps they liked best.

180 preferred Smurfs
270 preferred Moulden Wonder
50 preferred various other makes
100 expressed no preference.

(a) Display this information in a bar chart. (*4*)

(b) What percentage preferred Moulden Wonder? (*3*)

(c) Write the ratio of the numbers of those preferring Smurfs: all others in the survey in its lowest terms. (*4*)

Those who preferred Smurfs were also asked which flavour they liked best.

105 chose Malt and Vinegar
50 chose Pokey Bacon
25 chose Peas and Onion.

(d) Display this information in a pie chart. (*6*)

3. 1 kg is equal to 2.2 pounds (lb).

(a) Make out a graph suitable for converting masses of up to 10 kg to pounds. (*5*)

(b) Use your graph to convert

(i) 8.2 kg to pounds (*2*)

(ii) 1 stone (14 lb) to kg. (*2*)

(c) I can buy biscuits in 500 g packets. How many packets will I need, to have 20 lb of biscuits? (*5*)

(d) How much less is 14 lb than 8 kg? Answer in kg. (*3*)

Exercise 3.9

1. A swimming pool is made in the shape of two cuboids put together as shown in Figure 1.

(a) Find the area of the figure ABCDEF. *(4)*

(b) Find how much water would be required to fill the pool,
(**i**) in m^3 (**ii**) in litres. *(4)*

(c) What fraction of the pool has been filled when the water is 50 cm from the top of the pool? *(4)*

(d) How long would it take to fill the pool to this level if a pump were able to deliver 2500 litres a minute to the pool? *(5)*

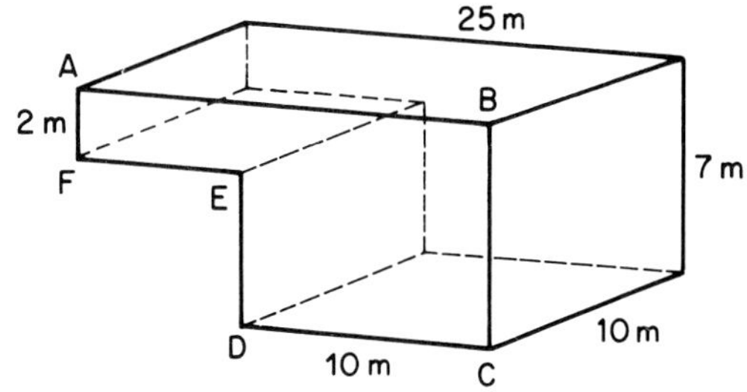

Figure 1

2. A stretch of straight coastline runs due East. Two coastguard stations, A and B, lie 10 km apart, right on the coast, A being due West of B. A ship S is spotted on a bearing of 043° from A and at the same time on a bearing of 321° from B.

(a) Find, by accurate drawing, the distance of the ship from A. *(8)*

(b) Construct, using ruler and compasses only, the perpendicular from S to the coast. Measure this distance and hence give the distance of the ship from the coast. *(5)*

(c) Letting the nearest point on the coast be X, find the distance from B to X. *(3)*

3.

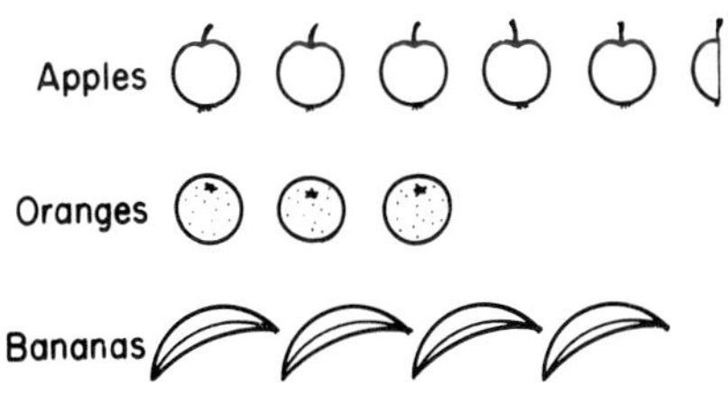

Figure 2

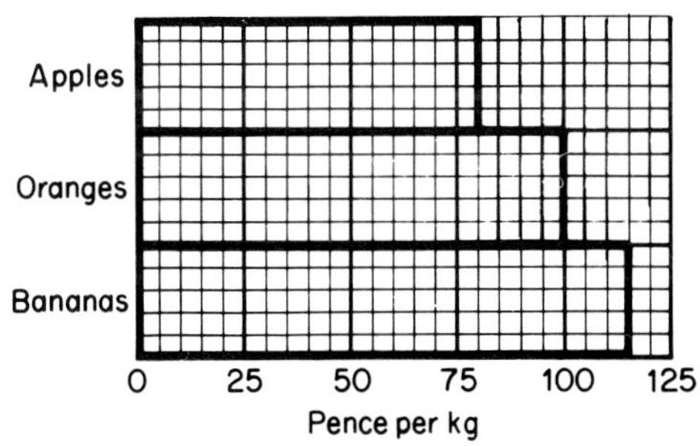

Figure 3

The pictogram (Figure 2) shows the weekly fruit consumption of the Smith family. The bar chart (Figure 3) shows the price per kg of each kind of fruit.

(a) Use this information to calculate

(**i**) the amount the family spends on each kind of fruit *(6)*

(**ii**) the average cost to the family of each kg of fruit it buys. *(4)*

(b) Display the information you calculated in (a) (i) as a pie chart. *(7)*

Exercise 3.10

1. A train has 3 coaches and 10 trucks.

(a) If the length of a coach is $5x$ metres, express the length of 2 coaches in terms of x. *(3)*

(b) The length of two coaches is equal to the length of five trucks. Express the length of a truck in terms of x. *(4)*

(c) The length of the train (without the engine) is 140 m. Form an equation and use it to find the length of a coach. *(6)*

(d) Find the length of a truck. *(3)*

2. Brian takes three magazines each week. *Computer Sense* costs 80 pence, *Electronics World* costs 56 pence.

(a) What is the cost of *Computer Sense* for 21 weeks? *(3)*

(b) What is the cost of *Electronics World* for 21 weeks? *(3)*

(c) If he pays £43.68 for magazines for the 21 weeks, how much does the third magazine, *Sports Records* cost for 21 weeks? *(4)*

(d) How much per copy is *Sports Records*? *(3)*

(e) If the cost of the magazines rose by 10%, what would Brian have to pay per week for the three magazines together, to the nearest penny? *(4)*

3. (a) The graph on the right shows two trains, A and B, travelling in opposite directions between two cities, X and Y, and stopping at stations on the way.

(i) After how long, and how far from X, did the two trains pass? *(4)*

(ii) How long did B wait at the intermediate station? *(2)*

(iii) What was B's average speed on the first part of its run? *(4)*

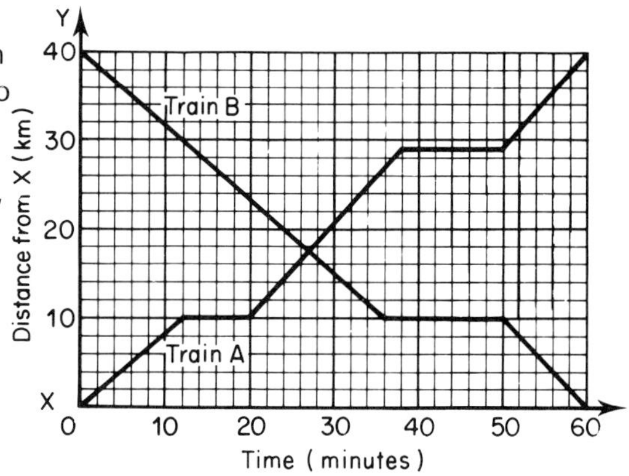

Figure 1

(b) **(i)** Find the average age of four boys, whose ages in April are 13 years 2 months, 12 years 11 months, 13 years 5 months and 13 years 2 months. *(5)*

(ii) What will be their average age in September of the same year? *(2)*

Exercise 3.11

1. A boy has 25 wooden cubes, all of the same size. He uses them to form a square of nine bricks standing on a square of 16 bricks, as shown in Figure 1. The edge of a brick is 2 cm.
   (a) What is the total surface area of the outside of the pile, including the area of the faces touching the floor? *(12)*
   (b) What is the total volume of the shape? *(5)*

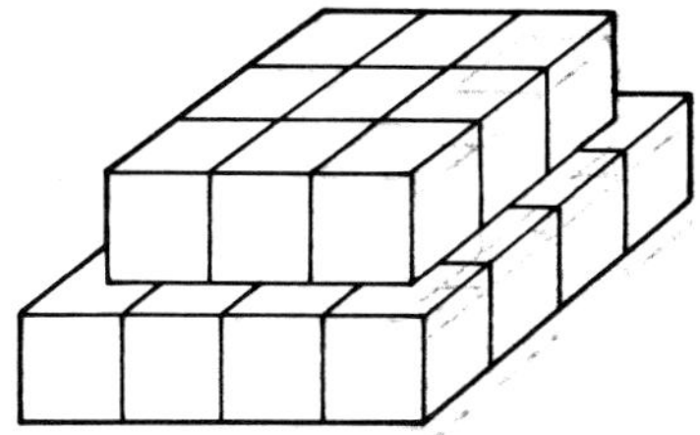

Figure 1

2. (a) Construct a triangle ABC with AB = 10.3 cm, BC = 8.8 cm and angle B = 40°. Measure AC. *(6)*
   (b) Bisect the lines AB and BC at right angles. Let these lines meet at a point X. Measure AX. *(5)*
   (c) With centre X and radius AX, draw a circle through the points A, B and C. *(2)*
   (d) Shade in carefully all points in the triangle nearer to B than A and nearer to C than B. *(4)*
3. Trevor travels x km at 18 km/h and three times as far at 12 km/h.
   (a) How far did he travel altogether, in terms of x? *(3)*
   (b) What was the total time for the journey, in terms of x? *(5)*
   (c) Given that the total time was 1 hour 50 minutes, form an equation and use it to find the value of x. *(8)*

Exercise 3.12

1. (a) Draw a sketch of a pentagon using pencil and ruler only. (*3*)
   (b) By drawing as many diagonals as possible from one of the vertices (corners), divide the pentagon into triangles. (*4*)
   (c) Using the drawing in part (b), and the number of triangles, work out the sum of the interior angles of the pentagon. (*5*)
   (d) What would each interior angle be, if the pentagon were regular? (*3*)
   (e) What would each exterior angle be, if the pentagon were regular? (*2*)

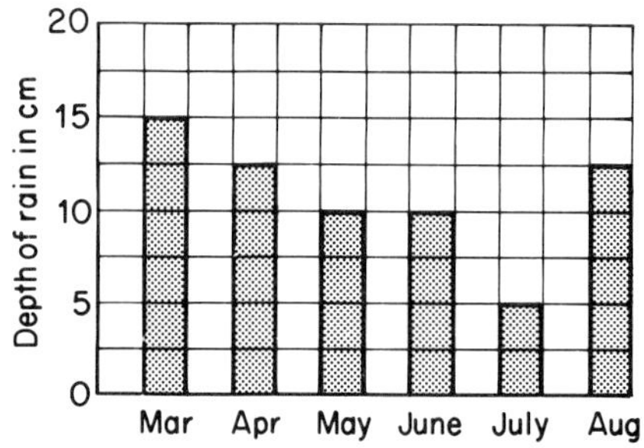

Figure 1

2. Figure 1 shows the total rainfall per month for the six months from March to August on the Costa Mathematica.
   (a) How many times as much rain fell in the wettest month as in the driest? (*3*)
   (b) What was the total rainfall for the six months? (*6*)
   (c) What fraction of the total fell in June and July? Give your answer in its lowest terms. (*4*)
   (d) Calculate the average *weekly* rainfall for the six months, supposing that there are exactly 26 weeks in the six-month period. (*4*)
3. A man works 48 hours each week, at a salary of £125 per week.
   (a) If his employer decides to reduce his hours each week in the ratio 5 : 6, how many hours will he now work? (*5*)
   (b) His salary per week is increased in the ratio 26 : 25. How much does he now earn per week? (*5*)
   (c) What does he now earn per hour? (*6*)